Wildlife Population Monitoring

Edited by Marco Ferretti

Published in London, United Kingdom

IntechOpen

Supporting open minds since 2005

Wildlife Population Monitoring
http://dx.doi.org/10.5772/intechopen.81350
Edited by Marco Ferretti

Contributors
Nicole Benjamin-Fink, Laurie Marker, Yogendra Shah, Dhan Kumar Pant, Yagya Raj Pandeya, Amir Sadaula, Rabin Kadariya, Erin Stukenholtz, Richard Stevens, Tirhas Hailu, Sean Childers, Charles Leatherwood, Lonnie Evans, Don Roulain, Dale Townsley, Marty Treider, R. Neal Platt, David Ray, John Zak, Guiming Zhang, Abiud Lucas Kaswamila, Augustino Mwakipesile, Magdalena Larska, Michał Krzysiak, Md. Akhter Hossain, Rajib Mahamud, Nikhil Chakma, Mohammed Hossain, Maria Dos Anjos Pires, Anabela Alves, Fernanda Seixas, Isabel Pires, Maria De Lurdes Pinto, Adelina Gama, Filipe Silva, Roberto Sargo, Madalena Vieira-Pinto, Alexandra Esteves, Estela Bastos, Leonor Orge, Carla Neves Machado, João Carlos Silva, Ana Paula Mendonça, Paula Tavares, Carla Lima, Paulo Carvalho, Ana Matos, Luis Figueira, Nuno Gonçalves-Anjo

First published in London, United Kingdom, 2019 by IntechOpen
IntechOpen is the global imprint of INTECHOPEN LIMITED, registered in England and Wales, registration number: 11086078, 7th floor, 10 Lower Thames Street, London, EC3R 6AF, United Kingdom
Printed in Croatia

British Library Cataloguing-in-Publication Data
A catalogue record for this book is available from the British Library

Additional hard and PDF copies can be obtained from orders@intechopen.com

Wildlife Population Monitoring
Edited by Marco Ferretti
p. cm.
Print ISBN 978-1-78984-169-5
Online ISBN 978-1-78984-170-1
eBook (PDF) ISBN 978-1-83880-496-1

For EU product safety concerns:
IN TECH d.o.o., Prolaz Marije Krucifikse Kozulić 3, 51000 Rijeka, Croatia, info@intechopen.com or visit our website at intechopen.com.

Meet the editor

Marco Ferretti obtained a PhD in Animal Production, Health and Food Hygiene in the Mediterranean Climate Countries from the University of Pisa (2010); a master's degree in Sciences and Management of Wildlife and Environmental Resources from the University of Florence (2006); and a bachelor's degree in Protection and Management of Wildlife Resources from the University of Florence (2001). From 1999 to 2008 he worked as a freelancer in the field of wildlife conservation and management, population dynamics, animal behavior, zoology, ecology, biology, agronomy, forestry, statistics, GIS, and wildlife and environmental law. From 2008 to 2015 Dr. Ferretti worked at the Wildlife Management office in the province of Pistoia. Since 2016 he has worked at the Wildlife Management Office in Tuscany.

Contents

Preface

Wildlife management has evolved in recent times. In past years, the scientific and technical world has tried to set up standardized monitoring methods for each species, to give answers to the question: Which and how many animals are we talking about?

The results of monitoring with these techniques, increasingly innovative thanks to new computer and digital tools, led to the discovery of problems related to the conservation of species and their habitats. These problems can manifest in cases of low numbers and/or residual areas, or in many cases, problems of interaction between animals and the population due to overabundance or damage to human activities.

The scientist and the technician therefore have increasingly had to deal with opposing situations: conservation of endangered species and mitigation of the impacts of wildlife on humans.

In these two contexts numerous factors come into play, including local interests from various stakeholders, national laws and international treaties, and perception of reality.

Nowadays, with a distorted use of the Internet and social networks, the risk of misperceptions surpassing the real facts is concrete.

Only a few roads to follow remain to the technician who works with fauna: combine the indispensable use of new technologies with a diligent and careful work in the field and continuous professional development.

This book deals with the monitoring of fauna, related diseases, and interactions with humans. It is intended to assist and support the professional worker in wildlife management

Marco Ferretti PhD.
Wildlife management office - Region of Tuscany, Italy

Section 1

Wildlife Management

Chapter 1

Wildlife Population Monitoring Study among Endangered Animals at Protected Areas in Nepal

Amir Sadaula, Yagya Raj Pandeya, Yogendra Shah, Dhan Kumar Pant and Rabin Kadariya

Abstract

Nepal is a small country located in South Asia which is geographically highly diverse and fairly rich in its variety of endangered wild animals. Conservation and monitoring of endangered wildlife is great challenging task in developing country like Nepal. Different body parts of wild animals are used as raw material for making pharmaceutical products, cosmetic, and other purpose; hence their value increased in the international market, as a result poaching and trading activities of endangered wild animals has been increased rapidly at protected areas in Nepal. This review will provide detailed information about different population monitoring techniques being applied for different endangered wild animals in protected areas of Nepal. Techniques like camera trapping, radio collar, noninvasive methods, mobile application, GPS, GIS, direct head count, etc. are commonly used for monitoring wildlife in Nepal. Since conservation of endangered wildlife species has become very much crucial, more advance technologies and social network analysis will be also used to determine the monitoring of wild animals at protected areas in Nepal.

Keywords: South Asia, conservation, protected areas

1. Introduction

Nepal is a small landlocked country located in Southeast Asia surrounded by India (South, East, and West) and China (North), which is geographically highly diverse and fairly rich in its variety of endangered flagship species, that is, wild animal and plants. The conservation of endangered wild animal species is a major challenge to government conservation of endangered wild animal species is a major threat as well as a challenge to government in developing countries particularly Nepal to protect the loss and degradation of natural habitats, poaching, and illegal trade activities of their crucial body parts for making pharmaceutical products, cosmetic, and other purpose [1]. Therefore, monitoring tools can be helpful to minimize the risk of possible human-animal conflict and also above described conditions especially to extinction of endangered wild animals in Nepal [2, 3].

However, the Government of Nepal has already successfully used the modern technologies such as radio collars and camera trap to monitor and conserve some endangered species like tigers and vulture at protected areas [4–7]. In the other hand, monitoring technologies have also several limitations, for example, camera trapping and radio collar require a long period of time as well as high costs [8]. Noninvasive genetic analysis from scat, urine, and hair samples have become widely used in wildlife research study, where molecular markers helps to identify the species, sex, and individuals [9, 10]. noninvasive genetic analysis from scat, urine, and hair samples have become widely used in wildlife research study, and additionally molecular markers can also help to identify the species, sex, and individuals [9, 10]. Noninvasive genetic analysis technique is more safe and avoids the potential harmful effects from direct contacts with endangered wild animal species while performing research at protected areas [11]. In Nepal, more than dozens of previous studies on endangered wild animal species have been successfully recognized by applying the noninvasive genetic analysis insight into population structure, mitochondrial (mt) DNA of cytochrome b and D-loop control region sequencing, effective population size, phylogenetic inference, and other modeling [9, 10, 12, 13] approach as described in **Table 1** [14–21].

Nepal has a total area of 147,181 square kilometers of which protected area mainly includes forest land area of different altitudes in terai, hills, and mountains. This possesses a multitude of landscape and maintains vast biodiversity in the Palearctic and Indo-Malayan ecozones. In Nepal, altitude ranges from 67 m in the southeastern Terai to 8848 m at the highest peak in the world "the Mount Everest." It has 11 bioclimatic zones which are ranging from lower tropical below 500 m to nival above 5000 m in the high Himalayas containing 9 terrestrial eco regions with 36 vegetation types. Nepal attains 10th rank in terms of richest flowering plant diversity in Asia having 1120 species of nonflowering plants and 5160 species of flowering plants. According to recording made by zoologists, Nepal has 181 mammal species, 844 bird species, 100 reptile species, 43 amphibian species, 185 fresh water fish species, and 635 butterfly species. For protection of biodiversity, the Government of Nepal has established 20 protected areas since 1973 AD consisting 10 national parks, 3 wildlife reserves, 6 conservation areas, and 1 hunting reserve. Suklaphata and Parsa wildlife reserves were upgraded to national parks in 2017 AD. Nine different Ramsar sites have been declared between 1988 and 2008.

Wild animal population is monitored for different reasons like to know the status of species whether it is in endangered or threatened state, biological interest or research purposes, game management, population being monitored to assess recovery or progress of any threatened species, biological diversity study, to know the effects of human management actions, land use patterns, etc. [7, 22, 23].

Very much careful consideration has to be done in collecting information about wild animals for monitoring wildlife population. Direct observation of animal and behavior observation is important for wildlife population monitoring. Tracks and dung of animal, thermal cameras or attaching monitoring devices like collars, leg bands, and data loggers are also used for population monitoring [7].

The purpose of this review study is to highlight the different population monitoring techniques being applied in Nepal for different endangered wild animals and their application in the field condition. Based on these applied techniques, counting of animals and their status and further research and conservation strategy are made.

DOI: http://dx.doi.org/10.5772/intechopen.84365

Endangered species		Genetic techniques				Conservation location	Monitoring method	Reference
Common name	**Scientific name**	**Sample**	**Sample size**	**Genetically profile remarks**	**Genetic analysis**			
Snow leopard	*Pantheria unica*	Scats	71 putative scats	(i) 19 (27%) scats: genetically identified; 10 (53%) successfully genotype at 6 microsatellite loci. (ii) Total 9 individual snow leopards; 4 individual identified at SPNP (1 male and 3 females) and 5 (2 male and 3 females) at KCA.	PCR; mitochondrial cytochrome b gene/ carnivore specific/ species /sex/individual identification	(SPNP) (KCA)	Noninvasive genetic analysis	[14]
Bengal tiger	*Panthera tigris tigris*	Feceal sample/ tigerparts/ skin vespieces/ blood smear	(n = 770)120 (15 putative cases)	(i) All forensic samples (n = 15) were of tiger: 10 male and 5 females. (ii) Geo-source location for 9 of the 14 samples with 6 ± 8 nuclear DNA microsatellite loci (iii) 6 samples were assigned to BNP and 1 was an exact match to a female tiger previously profiled according to their fecal DNA reference database. (iii) 2 tiger samples were assigned to SWR (iv) 1 from CNP	PCR; mt DNA cytochrome-b (CYT-B)/ species/sex/individual identification	CNP BNP SWR PWR	Noninvasive genetic analysis/ genetic structure analysis/ Geo-source assignment of unknown tigers/Molecular forensic techniques	[15]
Himalayan black bear	*Ursus thibetanus laniger*	Fresh bear feces/hair sample	n = 126 fecal samples n = 21 hair samples	8 microsatellite loci by using 147 samples: 60 individual identified.	PCR: mt control region (CR) D-loop/ cytochrome-b/whole mitogenome sequencing	ACA	Non-invasive genetic diversity/microsatellite genotyping/mitochondrial DNA sequencing/phylogenetic inference	[16]

Endangered species		**Genetic techniques**				**Conservation location**	**Monitoring method**	**Reference**
Common name	**Scientific name**	**Sample**	**Sample size**	**Genetically profile remarks**	**Genetic analysis**			
Hanuman langur	*Semnopithecus entellus*	Fecal sample	67 non-invasive fecal sample	(i) DNA sequences; successfully; 67 non-invasively collected fecal samples belonging to 18 wild Hanuman langur troops covering the entire distribution range of the species in Nepal. (ii) identified 37 haplotypes from the concatenated CR + CYTB (2230 bp) sequences	PCR: mt DNA control region (R = 1090 bp), cytochrome B (CYTB=1140 bp) sequences	KRS GRS KMRS	Non-invasive genetic diversity/population genetic structure/ molecular analysis/ paleodistribution modeling	[17]
Assam macaques	*Macaca assamensis*	Fecal sample	227 fecal samples collected from 39 wild troops	The mt DNA fragment complete control region (1121 bp): recovered from 208 fecal samples defining into 54 haplotye.	PCR: mt DNAcomplete control region (1121 bp)		Non-invasive genetic analysis/population genetic structure/paleodistribution constrcution	[18]
Tiger	*Panthera tigris tigris*	Fecal sample (Scat)	770 scat samples	(i).770 scat samples; 412 (57%) (ii) Out of 10 microsatellite loci; 8 markers identify 78 individual tigers (iii) Sex was genetically identified 353 scat samples; 255 samples (male) and 98 female	PCR:mt DNA cytochrome B fragment (162 bp)/species/ sex/individual identification	SuNP BNP BaNP CNP PNP	Non-invasive genetic analysis/genetic structure/ population structure/ genetic variation/ contemporay gene flow/ potential population	[19]
Snow leopard Himalayan Wolf	*Panthera unica* *Canis lupus chanco*	Scat samples	N = 573 leopard scat samples N = 236 wolf scat samples	N = 182 Wolf = 57	PCR:mt DNA cytochrome B/ species/sex/ individual identification	ACA MCA	Non-invasive genetic analysis	[20]

DOI: http://dx.doi.org/10.5772/intechopen.84365

Endangered species		Genetic techniques				Conservation location	Monitoring method	Reference
Common name	Scientific name	Sample	Sample size	Genetically profile remarks	Genetic analysis			
Himalayan Wolf	*Canis lupus chanco*	Fecal samples	N = 6 fecal samples	Out 6 fecal samples; 5 samples were successfully ampfified	PCR: mt DNA control region CR locus (220 bp)	ACA	Non-invaasive genetic analysis/molecular analysis/phyolgenetic analysis	[21]

SPNP, Shey Phoksundo National Park; KCA, Kangchanjunga Conservation Area; BNP, Bardia National Park; CNP, Chitwan National Park; PWR, Parsa WildlifeReserve; SWR, Suklaphanta Wildlife Reserve; ACA, Annapurna Conservation Area; KRS, Koshi River System; GRS, Gandaki River System; KMRS, Karnali-Mahakali River System; SuNP, Suklaphanta National Park; PNP, Paras National Park; MCA, Manasula Conservation Area.

Table 1.
Non-invasive genetic analysis studies were identified among the endangered wild animals species at protected areas in Nepal from 2011-2018.

2. Monitoring systems

Although conservation practice has come since long time in Nepal, only few biodiversity monitoring systems exist. Among existed system, the address is only for some mammal species of Terai region and one species of mountain (snow leopard). In monitoring process, repeated measurements are taken, and comparison is made to understand the cause of change.

Therefore, biodiversity monitoring includes carrying out repeated survey works to find out the size and extent of population of certain species. It also includes quality of habitat of certain species analysis results to find out trend and rate of change of species being monitored. By using standard methods and different field protocols, data is collected, and it is analyzed to determine the rate of progress. In the field data collection procedure ecologists and managers will be assisted by data analysis and GIS techniques.

GIS allows team to identify areas where they live. By overlying within a GIS environment, areas which match all criteria can be identified within minutes. This saves much fieldwork time, especially on hills and mountains where most wildlife habitat areas need to be accessed only on foot. Another important contribution made by GIS is in locating sample plots, which is crucial for validating datasets and results obtained from them. By the use of GIS, sample plot location can be identified either randomly or systematically. Thus identified sample plots can be overlaid within the three-dimensional GIS model, which will be helpful in providing exact impression of real field location. Location of these plots can be extracted from the GIS and will be identified in the field using GPS.

GIS uses uniform addressing format in the form of latitude, longitude, and altitude which is another advantage. Geostatistical techniques offer interpolation tools that take into consideration continuously changing spatial variables, which cannot be performed with general statistics. Such interpolation tools are useful for generating stoking density, especially of floral biodiversity [23–25].

2.1 Tiger monitoring (*Panthera tigris*) in Nepal

Abundance (N) and density (D) are important population parameters estimated for tiger monitoring programs. Photographic capture-recapture through camera trap is a most reliable technique for estimating tiger population as it being elusive species and the use of the unique identification patterns on each individual. Capture-recapture models provide a statistically robust framework to estimate species abundance, particularly when a population is said to be closed to births, deaths, immigration, or emigration during the survey period [26]. For camera trapping protected area or forest is divided into grids of 2 km × 2 km area, and a pair of cameras is placed in each grid cells. Field sampling design is determined by the size of the survey area and availability of logistics like camera number and personnel. Camera trapping was conducted in shifting blocks as described by Royle et al. [25] in each protected area and surrounding forests, which are divided into several blocks. Stations for camera trap are determined through prior sign survey of area like pugmarks, scrapes, and scat. In each station, two cameras are placed facing each other at a height of 45 cm above ground and were mounted on trees or posts on either side of a forest trail or road, with a distance of 6–8 m between the two cameras. In every grid cameras are placed for the standard sampling period of 15 nights. Each camera and memory card is given a unique identification number for data recording and maintenance purposes. Camera traps were checked every second day to ensure they were operating effectively.

Tiger habitat occupancy surveys are conducted across all potential tiger habitats. A grid cell each (15 km × 15 km) was laid across the survey area. Each grid cell is further divided into sub-cells. To include an element of randomness in the spatial distribution of survey routes, one sub-cell per grid cell coded as tiger habitat was randomly selected prior to the survey [27]. The number of spatial replicates per grid cell (i.e. km walked) was proportional to the percentage of tiger habitat [27]. For grid cells with 100% tiger habitat, we sampled 40 km in the cell touching random grid in every sampling route. Each contiguous 1 km segment was considered as a "spatial replicate" [28–30]. The field team walked along trails, roads, ridgelines, and river and stream beds in selected sub-cells, searching for tiger signs (scats, scrapes, pugmarks, kills, and urination sites), prey signs (dung, footprints, calls, and sightings), and human disturbance (wood cutting, lopping, grazing, poaching, etc.) following high probability tiger sign areas [28]. Observations were recorded for every 100 m section of the transect walk (**Figure 1**).

2.2 Rhino (*Rhinoceros unicornis*) monitoring in Nepal

The greater one-horned rhino was once believed widespread throughout the northern floodplains and nearby foothills of the Indian subcontinent between Indo-Myanmar border in the east and Sindh River basin, Pakistan, in the west; now greater one-horned rhinoceros are currently restricted in few protected areas in northeastern India and lowland terai of Nepal. Different techniques used for rhino monitoring in Nepal are mentioned below.

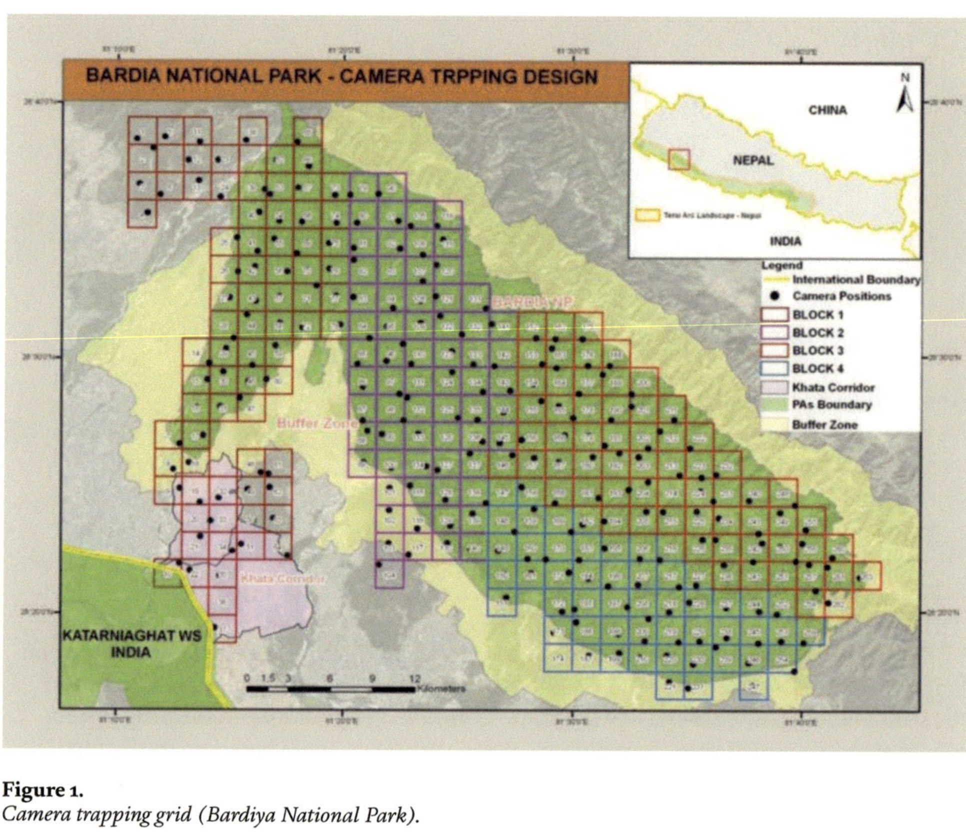

Figure 1.
Camera trapping grid (Bardiya National Park).

2.2.1 Rhino census

According to different physical demarcation like rivers and other field knowledge, potential rhino habitats (both inside and outside the protected areas) are divided into blocks. Reconnaissance surveys are also conducted in the more difficult areas to refine the blocks and plan for the sweeping operation. Where necessary, tall grasses are cut and burnt to improve visibility to observe animals. Elephant is an essential component for this census. Elephants are lined up over 5 km long with visible gap between the elephants and moved parallel along transects marked on the map to sweep individual blocks. Experienced staffs with camera, radio communication, GPS receivers, and maps are assigned to manage sections of the sweep to coordinate the operation. Radio communication is essential to maintain distance of approximately 50 m in dense forest and 100–200 m in open grassland so that rhino will not be missed. Maps of area to be covered on each census day are provided to each observer for navigation purposes. The maps in conjunction with the GPS receivers are extremely useful to ensure groups of elephants did not break from the census line particularly in thick forests (**Figure 2**).

A rhino data recording booklet is designed to standardize the accurate recording of rhino identification features and demographic data. The booklet convenient to fit in a shirt pocket is put together to form a convenient way of recording all of the details about each greater one-horned rhino. The observers are trained to look for and draw/note any features such as horn shape, ear tears, skin folds, deformities, epidermal knobs, body scars, or tail shape that would make a rhino distinguishable. They are trained to correctly fill in the following information:

- Date and time
- GPS location
- Standardized age class (adult, subadult, or calf)
- Sex (male, female, or unidentified sex)
- Habitat type (tall grassland, short grassland, Sal forest, riverine forest, wetland, other)
- Distinct rhino identification features (ears and horn, body and tail features)
- Body condition (good, average, poor)
- Group composition
- Activity

2.2.2 ID-based rhino monitoring

ID-based rhino monitoring is carried out in low population density area where population can differentiate easily and manually track and monitor individual animals and their movements. Intensive ID-based rhino monitoring is a simple monitoring system where each individual rhino is assigned a particular ID or name based on its distinct physical appearance and body features. All potential rhino habitats are divided into several blocks for ID-based rhino monitoring. The basic equipment required for rhino monitoring comprises cameras, binoculars, GPS, maps, and data forms. Based on the

DOI: http://dx.doi.org/10.5772/intechopen.84365

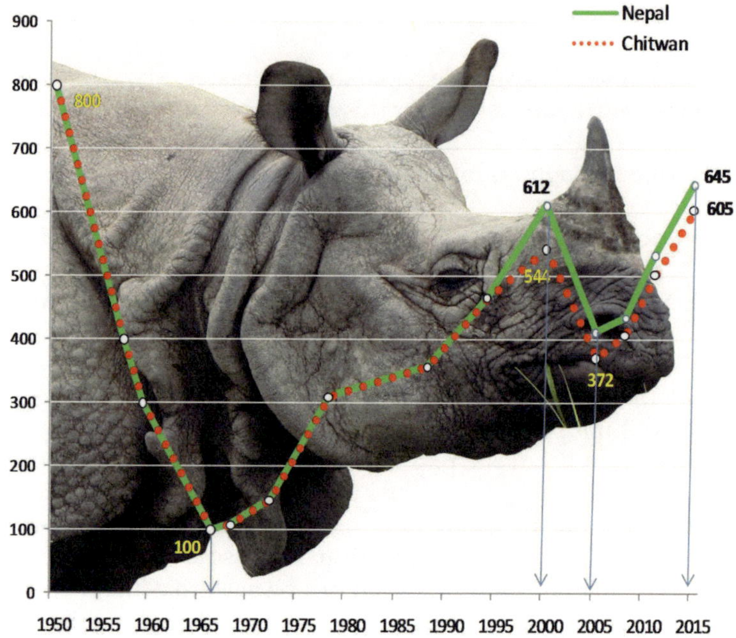

Figure 2.
Trend of Rhino Population in Nepal and Chitwan National Park.

field information of an individual rhino, a master ID file is prepared which is an invaluable tool for monitoring the rhino population. Rhino identification master files are used to control quality of rhino sighting data recorded in field sighting forms and trace the record of each rhino. The files also help to capture and transfer the knowledge/skills of the highly experienced key observers for future generation. The photographic sequences kept in the files enable to assess the changes in the animals over time and allow development of guidelines on horn and body size appearance with rhino age [31]. This also helps to monitor the animals for anti-poaching control. It is therefore essential that the information in the files is accurate and kept up-to-date by the data controllers with change in information like calving number, inter-calving intervals, and number of calves.

Once rhinos are individually identifiable, their details are maintained in population databases which assist greatly in ensuring that information can be derived to meet the needs both at park and national (metapopulation management) levels [31].

2.2.3 Monitoring rhinos through radio tracking

VHF, GPS, and satellite radio transmitter neck collars are modern radio-tracking technology which facilitates the monitoring of rhino. However, these are considerably more expensive. The transmitters can also include mortality sensors that change the frequency of the signal after a predetermined period of immobility.

The radio-tracking technology can be cost effective in certain situations such as:

- To monitor severe poaching activity.
- To monitor post-release adaptation of translocated rhinos to assess the adaptation in new habitat.
- Where there is insufficient monitoring capacity to ensure regular sightings through recognition of identity features

2.2.4 Mark-recapture population estimation

Mark-recapture methods provide population estimates. Mark-recapture population estimation software can be used for this purpose provided:

- Rhino sighting throughout subpopulation over a period of time.
- Equal attention been paid to monitoring both identifiable and "clean" rhinos.
- There are enough sightings of adults and independent subadult rhinos.

This technique of estimation is useful in populations where not all animals are individually identifiable and information are collected by nonspecialized teams of anti-poaching patrols and other staff on an ongoing ad hoc basis. Additional knowledge about the population will be derived (mortality, introductions, and removals in a population and where known calves have become independent of their mothers), are obtained through estimation process, and violates classical mark-recapture estimation. In these situations, periodic discrete surveys of a rhino subpopulation can be used to generate population estimates, using basic methods of mark-recapture population monitoring. Such estimates may have a lower degree of accuracy and precision than those that would be derived through the more continuous monitoring but will nonetheless be useful.

2.3 Red panda (*Ailurus fulgens)* monitoring in Nepal

First national red panda population count was made in 2016 by Red Panda Network in collaboration with the Government of Nepal. During this count program, habitat survey was also conducted. This study was conducted in 35 different districts of Nepal, and presence of red panda population was obtained in 23 districts of the country including 7 protected areas. This study provided information on deforestation level, habitat quality of red panda, and effect of climate change in their range.

A mobile application is being used to monitor movement of red panda, in eastern hill districts, including Panchthar, Ilam, and Taplejung. Wildlife activists who have received the phones have been entrusted with recording situation, speed, appearance, direction, and signs of the red panda and uploading them into the network through the Internet. GPS, camera, and mobile apps are used to monitor the red panda in Nepal [32–35].

2.4 Elephant (*Elephas maximus)* monitoring in Nepal

Photographic capture-recapture (CR) methods are sometimes adopted to estimate elephant population densities. However, this approach faces many functional challenges posed by difficulties in temporal sampling and unreliable identification of individuals on variable traits. Data of elephants were collected by photograph from roads and along water hole.

If an elephant is seen, it is recorded as fresh encounter, and its detailed description is made. Every elephant's morphological characteristics are classified as variable traits (holes and tears), fixed traits (tusk characteristics and ear fold shape), and tail traits (tail length, tail brush shape). Height of animal and age are recorded. Each animal is marked with different identity code [36].

2.5 Snow leopard (*Panthera uncia*) monitoring in Nepal

Camera trapping is also applied for the monitoring of snow leopard. Beside this satellite radio collaring in Kanchenjunga Conservation Area of eastern Nepal has been started since 2013 in collaboration with WWF Nepal. Both Conservation Committee and citizen scientists are trained in GPS handling and camera trapping and monitoring of snow leopards and their prey. The use of advanced technology in wildlife research such as noninvasive genetic analysis, camera trap surveys, and GPS-satellite telemetry presents better opportunity to conduct ecology and behavioral studies of snow leopards and their prey and habitats [24].

2.6 Gharial (*Gavialis gangeticus*) monitoring in Nepal

Gharial population monitoring is carried by sweep operation in the river system with potential habitat of gharial. River system is divided into several blocks. Each block is observed by a team of observers on boat. They will move along the river system and record all presence of gharial along with the activity of them. Monitoring is carried out continuously for 3 days. Average number is calculated with a record of 3 days.

A study on population status and distribution of Gharial in Nepal was commissioned by the Department of National Parks and Wildlife Conservation and WWF Nepal in 2008 aiming to update the existing status and distribution pattern of gharial and also to assess the threats in gharial conservation in Nepal. Data were collected from direct count by the use of opportunistic search method and questionnaire survey from Karnali and Babai Rivers of western Nepal, Narayani and Rapti Rivers of central Nepal, and Koshi River of eastern Nepal [37].

2.7 Swamp deer (*Cervus duvauceli*) monitoring in Nepal

Swamp deer count is carried out at the reserve by Suklaphanta Wildlife Reserve in partnership with Terai Arc Landscape Program and Suklaphanta Conservation Program, NTNC. The reserve is home to the world's largest herd of swamp deer. The count was based on direct head count of animals. Swamp deer is a protected species of Nepal and listed in Appendix I of CITES.

2.8 Gaur (*Bos gaurus*) monitoring in Nepal

For the census, protected area and surrounding forest are divided into several blocks. Sex, age, location, group size, habitat, and time are recorded. An elephant is deployed along with the observers; they will ride in the back of the elephant and count the individual head. Global positioning system (GPS), digital camera, binocular, maps, and communication sets were used for the census purpose. In Nepal, the Gaur is listed as protected animal under National Parks and Wildlife Conservation Act 2029, as vulnerable in IUCN Red Data Book, and is listed in Appendix I of CITES.

2.9 Musk deer (*Moschus chrysogaster*) monitoring and conservation status in Nepal

Musk deer is one of the genera called primitive deer that produced musk and high value of raw materials for making cosmetic as well as pharmaceutical products which is why musk can sell up to US $45,000 per kilogram (2.2 pounds) on the international market. However, musk is produced in the gland of males; later it can be extracted from live animals; for that purpose humans kill the musk deer to remove the entire sac having yields only about 25 g (1/40 of a kilogram) composed of the brown waxy substance [38, 39]. Due to increased price of musk deer on the international market, activities of poaching and trade of musk deer have been reported to rapidly increase in the Himalayan region of Nepal [40]. The study from neighboring country also revealed that poaching and trade of the species of musk is also a threat to conservation in Uttarakhand state of India [41]. The population of musk deer rapidly decrease due to many reasons such as poaching for musk pod, habitat overlapping, excess harvesting of forest resources like grass, shrubs, and timber for the human benefits, causing a serious problem for the survival of musk deer in the Himalayan region of Nepal [42, 43]. Grazing of livestock within protected areas might also be a possible risk and grazing of livestock within protected areas might also be a possible risk and threat to musk deer for no availability of enough diet to survive and high chance of disease transmission from livestock to musk deer and vice versa [40].

However, there was only limited information on musk deer and lack of detailed study on the distribution as well as threat in few protected areas in Nepal. Noninvasive genetic analysis was carried out to know information on the population status, distribution, habitat, seasonal diet, altitude range, survival life span, genetic diversity, and phylogenetic analysis of musk deer in Nepal.

3. Conclusion

Although Nepal is a small country, it has a vast biodiversity found from very low land to the highest mountain in the world. Conservation has come a long way in Nepal; few biodiversity monitoring systems exist and do address for some large species like tiger, greater one-horned rhino, elephant, crocodile, swamp deer, musk deer, snow leopard, red panda, gaur, etc. This review provides information on different population monitoring techniques that have been used in Nepal for endangered wildlife species. In Nepal, wildlife population is basically monitored to assess progress or recovery of endangered or threatened species and research purposes and to know the effect of human management actions. Different methods such as directly observing animals and their behavior in situ, looking for signs like tracks and dung of animals, radar, thermal cameras, capture/mark/release of animals, or attaching or implanting monitoring devices (e.g., collars, leg bands, back packs, or data loggers) and noninvasive methods using fecal, hair, etc. are used for population monitoring. In Nepal population monitoring is being done for few wildlife species only although it has many species. So, population monitoring of other species is also important for making proper conservation plan. In developed countries more advance monitoring technologies are being developed and adapted too. So it needs to collaborate with different biodiversity conservation organizations to work together and share technology and make monitoring more comprehensive. This present review study will provide strong recommendations to community persons, leaders, conservation NGO/INGO, and government bodies to prepare the future action plan strategies about the conservation and monitoring of flagship endangered wild animal species at protected areas in Nepal. It is important to know further detailed information on population status, distribution, habitat, seasonal diet, altitude range, survival life span, genetic diversity, and phylogenetic inferences of each particularly endangered wild animal found at protected areas in Nepal.

DOI: http://dx.doi.org/10.5772/intechopen.84365

Acknowledgements

National Trust for Nature Conservation, Chitwan, Nepal.

Conflict of interest

None.

Author details

Amir Sadaula[1†], Yagya Raj Pandeya[2†], Yogendra Shah[3,4*†], Dhan Kumar Pant[4] and Rabin Kadariya[5]

1 National Trust for Nature Conservation, Biodiversity Conservation Center, Chitwan, Nepal

2 Nepal Agricultural Research Council (NARC), National Cattle Research Program, Chitwan, Nepal

3 Everest International Clinic and Research Center, Kathmandu, Nepal

4 National Zoonoses and Food Hygiene Research Centre, Kathmandu, Nepal

5 Laboratory of Wildlife Biology and Medicine, Graduate School of Veterinary Medicine, Hokkaido University, Japan

*Address all correspondence to: yogendra90@hotmail.com

†Equally contribution authors.

References

[1] WWF International, ZSL. Living Planet Report 2014. ZSL: WWF; 2014

[2] Latham ADM, Latham MC, Anderson DP, Cruz J, Herries D, Hebblewhite M. The GPS craze: Six questions to address before deciding to deploy GPS technology on wildlife. New Zealand Journal of Ecology. 2015;**39**(1):143-152

[3] Kay R, Crofoot MC, Jetz W, Wikelski M. Terrestrial animal tracking as an eye on life and planet. Science. 2015;**348**(6240):2478

[4] WWF-Nepal. Nepal-India trans-boundary efforts for tiger conservation intensify as Chitwan sees a rise in tigernumbers. 2010. Available at: http://www.wwfnepal.org/?194368/Nepal-India-trans-boundary-efforts-for-tiger-conservation [Accessed: February 16, 2015]

[5] Government of Nepal. Terai Arc Landscape—Ministry of Forests and Soil Conservation Strategic Plan. Kathmandu: WWF Nepal; 2004

[6] Marvin DC, Koh LP, Lynam AJ, Wich S, Davies AB, Krishnamurthy R, et al. Integrating technologies for scalable ecology and conservation. Global Ecology and Conservation. 2016;7:262-275

[7] Shrestha Y, Lapeyre R. Modern wildlife monitoring technologies: Conservationists versus communities? A case study: The Terai-Arc landscape, Nepal. Conservation and Society. 2018;**16**(1):91-101

[8] Jackson RM, Roe JD, Wangchuk R, Hunter DO. Estimating snow leopard population abundance using photography and capture-recapture techniques. Wildlife Society Bulletin. 2006;**32**:772-781

[9] Janecka JE, Jackson R, Yuquang Z, Diqiang L, Munkhtsog B, Buckley-Beason V, et al. Population monitoring of snow leopards using noninvasive collection of scat samples: A pilot study. Animal Conservation. 2008;**11**:401-411

[10] Waits LP, Paetkau D. Noninvasive genetic sampling tools for wildlife biologist: A review of application and recommendations for accurate data collection. Journal of Wildlife Management. 2005;**69**:1419-1433

[11] Janecka JE, Munkhtsog B, Jackson RM, Mallon DP, Naranbaatar G, Gerelmaa K, et al. Comparison of noninvasive genetic and camera trapping techniques for surveying snow leopards. Journal of Mammalogy. 2011;**92**

[12] Ahlborn G, Jackson R. Marking in free-ranging snow leopards in west Nepal: A preliminary assessment. In: 5th International Snow Leopard Symposium; Dehradun, India. Freeman H. International Snow Leopard Trust and Wildlife Institute of India. 1988. pp. 25-49

[13] DeYoung RW, Honeycutt RL. The molecular toolbox: genetic techniques in wildlife ecology and management. Journal of Wildlife Management. 2005;**69**:1362-1384

[14] Karmacharya D, Thapa K, shrestha R, Dhakal M, Janecka JE, et al. Noninvasive genetic population survey of snow leopards (*Panthera uncia*) in Kangchenjunga conservation area, Shey Phoksundo National Park and surrounding buffer zones of Nepal. BMC Research Notes. 2011;**4**:516

[15] Karmacharya D, Sherchan AM, Dulal S, Manandhar P, Manandhar S, Joshi J, et al. Species, sex and geo-location identification of seized tiger (*Panthera tigris tigris*) parts in Nepal: A

molecular forensic approach. PLoS One. 2018;**13**(8):e0201639

[16] Kadariya R, Shimozuru M, Maldonado JE, Moustafa MAM, Sashika M, Tsubota T. High genetic diversity and distinct ancient lineage of Asiatic black bears revealed by non-invasive surveys in the Annapurna Conservation Area, Nepal. PLoS One. 2018;**13**(12):e020766

[17] Khanal L, Chalise MK, Wan T, Jiang X. Riverine barrier effects on population genetic structure of the Hanuman langur (*Semnopithecus entellus*) in the Nepal Himalaya. BMC Evolutionary Biology. 2018;**18**:159

[18] Khanal L, Chalise MK, He K, Acharya BK, Kawamoto Y, Jiang X. Mitochondrial DNA analyses and ecological niche modeling reveal post-LGM expansion of the Assam macaque (*Macaca assamensis*) in the foothills of Nepal Himalaya. American Journal of Primatology. 2018;**80**(3):e22748. DOI: 10.1002/ajp.22748

[19] Thapa K, Manandhar S, Bista M, Shakya J, Sah G, Dhakal M, et al. Assessment of genetic diversity, population structure, and gene flow of tigers (*Panthera tigris tigris*) across Nepal's Terai Arc Landscape. PLoS One. 2018;**13**(3):e0193495

[20] Chetri M, Odden M, Wegge P. Snow Leopard and Himalayan Wolf: Food habits and prey selection in the Central Himalayas, Nepal. PLoS ONE. 2017;**12**(2):e0170549. DOI: 10.1371/journal. pone.0170549

[21] Chetri M, Jhala YV, Jnawali SR, Subedi N, Dhakal M, Yumnam B. Ancient Himalayan wolf (*Canis lupus chanco*) lineage in Upper Mustang of the Annapurna Conservation Area, Nepal. ZooKeys. 2016;**582**:143-156. DOI: 10.3897/zookeys.582.5966

[22] Protected areas of Nepal [Internet] 2016. Available from https://en.wikipedia.org/wiki/Protected_areas_of_Nepal [Accessed: 2016-02-02]

[23] Witmer GW. Wildlife population monitoring: Some practical considerations. Wildlife Research. 2005;**32**(3):259-263

[24] DNPWC. Snow Leopard Conservation Action Plan (2017-2021). Kathmandu, Nepal: Department of National Parks and Wildlife Conservation; 2017

[25] Royle JA, Nichols JD, Karanth KU, Gopalaswamy M. A hierarchical model for estimating density in camera trap studies. Journal of Applied Ecology. 2009;**46**:118-127

[26] Karanth KU. Estimating tiger *Panthera tigris* populations from camera trapping data using capture-recapture models. Biological Conservation. 1995;**71**:333-338

[27] Karanth KU, Kumar NS, Srinivas V, Gopalaswamy A. In: Technical Support Team, Tigers Forever, editor. Revised Monitoring Framework for Tigers Forever: Panthera Sites. Bangalore, India: WCSIndia; 2008

[28] Barber-Meyer S, Jnawali S, Karki J, Khanal P, Lohani S, Long B, et al. Influence of prey depletion and human disturbance on tiger occupancy in Nepal. Journal of Zoology. 2013;**289**(1):10-18

[29] Hines JE, Nichols JD, Royle JA, MacKenzie DI, Gopalaswamy AM, Kumar NS, et al. Tigers on trails: Occupancy modeling for cluster sampling. Ecological Applications. 2010;**20**:1456-1466

[30] Dhakal M, Karki MK, Jnawali SR, Subedi NP, Malla NB, Lamichhane S, et al. Status of Tigers and Prey in Nepal. Kathmandu, Nepal: Department of National Parks and Wildlife Conservation; 2014

[31] DNPWC. The Status And Distribution of The Greater One-Horned Rhino in Nepal. Kathmandu, Nepal: Department of National Parks and Wildlife Conservation; 2009

[32] Bista D. Communities in frontline in red panda conservation eastern Nepal. Friends of Nature The Himalayan Naturalist. 2018;**1**(1):11-12. Available from: http://www.fonnepal.org/files/download/3Red%20panda%20conservation_the HimalayanNaturalist_Volume1Issue1May2018

[33] Tucker G, Bubb P, de Heer M, Miles L, Lawrence A, Bajracharya SB, et al. Guidelines for Biodiversity Assessment and Monitoring for Protected Areas. Kathmandu, Nepal: KMTNC; 2005. Available from: https://www.gim-international.com/content/article/biodiversity-surveying-and-monitoring?output=pdf

[34] Red Panda Field Survey and Protocol for Community Based Monitoring. Singha Durbar, Kathmandu: Government of Nepal Ministry of Forests and Soil Conservation; 2016

[35] Universities federation for animal welfare. Monitoring of wildlife population [Internet]. Available from: https://www.ufaw.org.uk/why-ufaws-work-is-important/monitoring-of-wildlife-populations

[36] Karanth KK. Developing simple and innovative techniques to monitor elephants. Conservation India. 2014. Available from: http://www.conservationindia.org/articles/developing-simple-and-innovative-techniques-to-monitor-elephants

[37] Khadka M, Kafley H, Thapaliya BP. Population status and distribution of Gharial (*Gavialis gangeticus*) in Nepal. In: Forum of Natural Resource Managers Report, Nepal; 2008

[38] Harris R. For musk deer and other wildlife in southern Qinghai, China. Mountain Research and Development. 1991;**11**(4):353-358

[39] Harris RB. Wildlife Conservation in China: Preserving the Habitat of China's Wild West. Armonk, New York, USA: M. E. Sharpe, Inc.; 2007

[40] Aryal A, Raubenheimer D, Subedi S, Kattel B. Spatial habitat overlap and habitat preference of Himalayan musk deer (*Moschus chrysogaster*) in Sagarmatha (Mt. Everest) National Park, Nepal. Current Research Journal of Biological Sciences. 2010;**2**:217-225

[41] Ilyas O. Status, habitat use and conservation of Alpine musk deer (*Moschus chrysogaster*) in Uttarakhand Himalayas, India. Journal of Applied Animal Research. 2015;**43**(1):83-91

[42] Aryal A, Subedi A. Conservation and potential habitat of Himalayan Musk deer *Moschus chrysogaster* in the protected areas of Nepal. International Journal of Conservation Sciences. 2011;**2**(2):127-141

[43] Yang Q , Meng X, Xia L, Feng Z. Conservation status and causes of decline of musk deer (*Moschus* spp) in China. Biological Conservation. 2003;**109**:333-342

Chapter 2

Integrating Citizen Science and GIS for Wildlife Habitat Assessment

Guiming Zhang

Abstract

With the rapid advancement and popularity of geospatial technologies such as location-aware smartphones, mobile maps, etc., average citizens nowadays can easily contribute georeferenced wildlife data (e.g., wildlife sightings). Due to the wide spread of human settlements and lengthy living histories of citizens in their local areas, citizen-contributed wildlife data could cover large geographic areas over long time spans. Citizen science thus provides great opportunities for collecting wildlife data of extensive spatiotemporal coverage for wildlife habitat assessment. However, citizen-contributed wildlife data may be subject to data quality issues, for example, imprecise spatial position and biased spatial coverage. These issues need to be accounted for when using citizen-contributed data for wildlife habitat assessment. Geovisualization and geospatial analysis capabilities provisioned by geographic information systems (GISs) can be adopted to tackle such data quality issues. This chapter offers an overview of citizen science as a means of collecting wildlife data, the roles of GIS to tackle the data quality issues, and the integration of citizen science and GIS for wildlife habitat assessment. A case study of habitat assessment for the black-and-white snub-nosed monkey (*Rhinopithecus bieti*) using *R. bieti* sightings elicited from local villagers in Yunnan, China, is presented as a demonstration.

Keywords: citizen observers, local villagers, wildlife sightings, geovisualization interview, habitat suitability mapping, data quality, black-and-white snub-nosed monkey (*Rhinopithecus bieti*)

1. Introduction

Habitats provide resources such as food, shelter, potential nesting sites, and mates for wildlife to achieve survival and reproduction [1]. Understanding the requirements or preferences of wildlife on their habitats and assessing the quality of wildlife habitat is of great importance for conservation biologists and conservation managers [2]. For example, wildlife habitat assessment supports conservation practices such as ex situ or reintroduction and restoration conservation, predicting risk of invasive species, systematic conservation planning, assessing threats, and setting conservation priorities [3–6].

One approach to assessing wildlife habitat quality is to predict wildlife habitat suitability maps indicating the spatial variation of habitat suitability [7]. Habitat suitability mapping is often carried out in a geographic information system (GIS) [8].

Conceptually, spatial prediction of wildlife habitat suitability requires GIS data layers characterizing the environmental conditions (environmental data) and knowledge on the relationship between wildlife habitat suitability and environmental conditions. Based on the relationship and the environmental conditions at a location (e.g., a pixel), the in situ habitat suitability can be inferred. Inferring habitat suitability at every location in the study area of interest results in a suitability map [7]. Such a habitat suitability map can then be used to assess the spatial variation of wildlife habitat quality and to support conservation. With the rapid development of geospatial technologies, environmental data for characterizing environmental conditions are becoming abundant and increasingly available [9, 10]. The key for wildlife habitat assessment through habitat suitability mapping therefore lies in obtaining knowledge on the relationship between wildlife habitat suitability and environmental conditions (environmental niche).

Data-driven approaches are most commonly adopted in deriving the relationship between wildlife habitat suitability and environmental conditions (environmental niche modeling) [7]. Data-driven approaches for environmental niche modeling require wildlife data indicating habitat use, for example, abundance data, presence and/or absence data. Wildlife data are overlaid with environmental data layers to extract the environmental conditions at locations where habitat use occurs. The relationship between wildlife habitat suitability and environmental conditions can then be derived through statistical analysis, machine learning, data mining, or other modeling techniques [7]. Thus, wildlife data become the key to deriving the relationship between habitat suitability and environmental conditions for mapping wildlife habitat suitability for wildlife habitat assessment.

Traditionally, wildlife data are collected using various techniques such as field observation, radio telemetry, infrared trapping cameras, and global positioning system (GPS) collars [11, 12]. Accurate wildlife data can be collected through these techniques, but admittedly these techniques are also somewhat expensive to deploy [13]. The high cost may prevent these techniques from being used in wildlife data collection, particularly for areas and projects with limited budget support. Besides, some of these techniques (e.g., field observation and GPS collars) are logistically difficult for areas with rugged terrains and limited accessibility [13]. Low-cost techniques such as trailing wildlife markings and interviewing local people about wildlife sightings through questionnaires are also used for wildlife data collection, but wildlife data collected with these techniques can be of low quality (e.g., inaccurate spatial location and/or time) [14, 15]. Cost-effective methods for collecting wildlife data of satisfactory quality are ideal for wildlife habitat assessment and sustainable conservation given that much of the world's biodiversity occurs in the world's poorest and remote countries [16].

Local residents were proven to be a cost-effective source of obtaining wildlife data [17, 18]. Many local residents, such as those living in remote rural areas and particularly those whose livelihoods are closely linked to ecosystem services (e.g., subsistence farmers, shepherds, and hunters), spend a great deal of time in the field. They encounter wildlife in its natural environment and, as a result, accumulate a rich knowledge about the wildlife habitat use. Wildlife data elicited from local residents at relatively low cost, although may be subject to data quality issues (e.g., data credibility, positional accuracy, spatial bias, etc.), could be used to support and sustain conservation programs with limited budget.

From a broader perspective, the increasing availability of citizen-contributed data accompanied by the advancements in GIS has created the opportunity to make full use of citizen science to address many real-world problems. On the one hand, citizen-contributed data have become increasingly available with the resurrected popularity of citizen science [19] and the emerging phenomenon of volunteered geographic information (VGI) [20]. One prominent example is the eBird citizen science project that is driven by bird watchers and documenting bird species across the globe [21]. On the other hand, the advancements in GIS capabilities (e.g., geovisualization and spatial analysis) have made it possible to accommodate the data quality issues associated with citizen-contributed data to make use of such data for scientific inquires [22, 23].

This chapter offers an overview of citizen science for wildlife data collection and its integration with GIS for wildlife habitat assessment. A case study of habitat assessment for the black-and-white snub-nosed monkey (*Rhinopithecus bieti*) using data contributed by local residents in Yunnan, China, is presented as an illustration.

2. Citizen science for wildlife data collection

2.1 Citizen science

The term *citizen science* was formally added to the Oxford English Dictionary only recently in 2014 [24], referring to "Scientific work undertaken by members of the general public, often in collaboration with or under the direction of professional scientists and scientific institutions" [25]. Nonetheless, citizen science has been practiced for centuries, long before *scientist* slowly became a profession throughout the seventeenth to nineteenth centuries [24]. For example [26], Benjamin Franklin (1706–1790), as a physicist, was famous for his discoveries and theories regarding electricity while he was also a printer, diplomat, and politician; Charles Darwin (1809–1888) as a biologist was best known for his contributions to the theories of evolution, but on the Beagle voyage, he was sailing as an unpaid companion, not as a professional scientist. Even after the *scientist-as-profession* paradigm has been well established, average citizens continue to engage in scientific work at various levels of involvement [27]: *contributory* where citizens mostly contribute to data collection, *collaborative* where citizens also participate in data analysis, and *co-created* where citizens get involved at all stages of the project including conceiving and designing the research. Exemplary long-running citizen science projects related to wildlife population monitoring are the Christmas Bird Count (CBC) established in 1900 [28] and the Breeding Bird Survey (BBS) established in 1965 [29] for monitoring bird species in North America. Data contributed by participants in such citizen projects are now supporting wildlife population trends monitoring [30] and decision-making in conservation [31].

The rapid advancements of geospatial information technologies in the last decade have greatly prompted the flourish of citizen science. Location-aware portable devices constantly connected to the Internet (e.g., GPS-enabled smart phones) are now commonplace. Average citizens thus can conveniently contribute georeferenced wildlife observations using such devices via social media, mobile map, citizen science project mobile apps, etc. [26, 32, 33]. From a geographic and GIS perspective, citizen science involving geospatial data generation (e.g., wildlife sightings with location information) is called "geographic citizen science" [34] and the georeferenced wildlife observations are a form of VGI [20, 34]. Due to the increasing availability of enabling technologies, millions of citizens across the world are participating in citizen science projects and many of them are contributing large volumes of wildlife observations on a daily basis. Interested readers can check out a wide range of ongoing citizen science projects (not limited to wildlife-related projects) at scistarter.com and search for specific projects by topic and/or location. As of the time of writing, searching projects at scistarter.com by the topic "Animals," "Birds," and "Insects & Pollinators" returned 382, 162, and 190 projects, respectively. As a prominent example, the eBird project [21, 35], launched in 2002 by the Ornithology Lab at Cornell University and the National Audubon Society, as of November 2016, has engaged over 330,000 bird watchers from more than 250 countries who have reported observations of over 10,300 bird species. As of June 2018, eBird has accumulated over 500 million bird observations in its global database; in recent years, there have been more than 100 million bird observations added to the database each year.

Wildlife data contributed by participants in such citizen science projects are a form of geospatial big data [36, 37]. Complex patterns can be discovered from such intensive data through visualizations, simulations, data mining, and various modeling techniques to provide valuable insight for forming concrete hypotheses about the underlying ecological, biological, and geographical processes that generated the observed data [37]. Thus, the abundance of citizen-contributed wildlife data has the potential of shifting research paradigm in biological, ecological, and geographical studies from the traditional hypothesis-driven approach to the emerging data-driven approach; for instance, scholars are promoting the idea of "data-intensive science" for biodiversity studies and "data-driven geography" [36–38].

2.2 The (dis)advantages of citizen science for collecting wildlife data

Citizen science has several advantages as an alternative mechanism for collecting wildlife data. Citizen-contributed data contain rich local information that spans a wide temporal spectrum because citizens, as local experts and sensors [20], have long been sensing and accumulating knowledge of their respective areas. Citizen science also has the potential to provide wildlife data over large areas, given that billions of networked human sensors are distributed across the globe. In addition, citizen science can provide timely updated wildlife data that are difficult to obtain and maintain through other techniques but can be easily elicited from citizens living in the local areas. Moreover, citizen-contributed data are much less expensive than traditional scientific data collection protocols (e.g., biological survey). In many cases, citizens contribute data purely voluntarily [20]. This low cost is of great practical significance in many real-world programs falling short of funding support.

Due to the above advantages of citizen science, it is possible to obtain timely updated wildlife data using citizen science over large areas. Citizen science thus has a great potential to support and sustain long-time wildlife population monitoring at large spatial scale (e.g., eBird) and provide wildlife data for wildlife habitat assessment.

In spite of the strengths, one should be aware of the shortcomings of the "citizen science" approach to wildlife data collection. For example, this approach cannot be used in areas with low population where sufficient local citizen observers/informants are lacking. It is also not good for collecting data on evasive animals with little contact with humans. Most importantly, there can be data quality issues associated with wildlife data contributed by volunteer citizens (i.e., non-professionals) which make the data challenging to standardize and analyze [17, 18]. The following sections detail some of the data quality issues, their implications for wildlife habitat assessment, and how GIS techniques (geovisualization, geospatial analysis, geocomputation, etc.) can be adopted to tackle the issues toward reducing the impact of such issues on wildlife habitat assessment.

2.3 The data quality issues of citizen-contributed wildlife data

The quality of citizen-contributed wildlife data is the major concern when using such data for wildlife habitat assessment. The average citizens engaged in citizen science projects are not well-trained professionals; their voluntary data collection

DOI: http://dx.doi.org/10.5772/intechopen.83681

actions are mostly constrained by internal commitment. Thus, citizen-contributed wildlife data may or may not be accurate [20, 39]. Three aspects of data quality are particularly relevant to the use of citizen-contributed wildlife data for wildlife habitat assessment: data creditability, positional accuracy, and spatial bias.

2.3.1 Data credibility

In order to be useful for wildlife habitat assessment, wildlife data (e.g., sightings) reported by citizen participants need to be credible, that is, provide ground truth wildlife observations. Data credibility is affected by the characteristics of both the wildlife and the citizen observers (e.g., local residents). On the one hand, local residents often only observe wildlife that is active in the daytime. The target wildlife should be easily recognizable to reduce misidentification given that local residents usually have no training on species identification [17, 40]. On the other hand, local resident knowledge of the target wildlife, age, length of residence, and formal education also influence data credibility [41]. For instance, performance in georeferencing tasks differs between novice and expert citizen participants [42]; there exists both between-observer differences [43] and within-observer differences (over time) [44] in BBS participant bird-counting skills.

Various methods have been developed for increasing the credibility of citizen-contributed wildlife data. Ref. [45] identified a total of 12 strategies that have been adopted by citizen science programs to increase their data credibility across different program stages including training and planning, data collection, and data analysis and program evaluation. As an example, eBird uses a two-part approach to assure data credibility during data entry [39]: automated data quality filters flag records for review based on observation date and geographic location; a flagged entry, once confirmed as legitimate by the observer, is then reviewed by a regional expert reviewer again.

2.3.2 Positional accuracy

Position of the wildlife data used for habitat suitability mapping needs to be accurate so that the locations can be used to accurately obtain the corresponding environmental conditions at these locations from environmental data layers. Insufficient positional accuracy of wildlife data leads to mismatch between the locations of wildlife habitat use and the corresponding environmental conditions, and thus degrades the accuracy of environmental niche modeling and habitat suitability mapping [46].

Nonetheless, it is also important to note that the impact of positional accuracy of wildlife data on habitat suitability mapping depends on the spatial resolution at which suitability mapping is conducted. Mapping at high spatial resolution (e.g., using environmental data of 30 m × 30 m grids) definitely requires wildlife data of high positional accuracy that is comparable to the spatial resolution of the environmental data so that values of the environmental conditions at these locations can be accurately extracted from environmental data layers. In contrast, for mapping at coarse spatial resolution (e.g., 1000 m × 1000 m grids), the absolute positional accuracy of wildlife data does not have to be very high as long as it is accurate enough relative to the spatial resolution of environmental data in use.

2.3.3 Spatial bias

Wildlife observations contributed by citizens are often concentrated more in some geographic areas than others (i.e., spatial bias) because observations made by citizens are opportunistic in nature [23]. Unlike well-designed sampling or survey schemes which allocate observation sites in a way such that the geographic space and/or the environmental space are well covered by the observation sites, spatial distribution of the observation efforts of citizen volunteers would be considered neither random nor regular in the sense of sampling or survey design. One example to demonstrate this is wildlife sightings elicited from local residents. Local residents are not intentionally tracking wildlife of interest. Instead, they typically spot the wildlife en route to doing something else. The routes on which local citizens spot wildlife would be considered neither random nor regular but "ad hoc" [23]. As a result, wildlife sightings elicited from local residents are usually concentrated in areas with higher route accessibility.

Such spatial bias in wildlife data has a significant impact on environmental niche modeling and habitat suitability mapping for wildlife habitat assessment. Due to the spatial bias, citizen-contributed wildlife data might not be representative of the actual wildlife habitat use. The relationship derived based on the wildlife data thus might not well represent the underlying environmental niche. Spatial bias in citizen-contributed wildlife data, if not appropriately accounted for, would adversely affect the accuracy of wildlife habitat suitability mapping [47–49].

3. The roles of GIS

GIS is the ideal tool for conducting wildlife habitat assessment as it involves geospatial data. Besides providing an integrated environment for managing and manipulating environmental data layers and georeferenced wildlife data, GIS can also offer capabilities to remedy or address some of the data quality issues associated with citizen contribute wildlife data. Firstly, geovisualization can be used to facilitate wildlife data elicitation from citizen participants and improve positional accuracy. Secondly, based on the cause of spatial bias, spatial analysis can be used to compensate for the biased coverage in observation efforts. Lastly, geospatial computation techniques can be employed to address the computational challenges arising from analyzing very large volumes of citizen-contributed wildlife data.

3.1 Geovisualization to improve positional accuracy

In general, positional accuracy of wildlife data largely depends on the availability of positioning technology. Wildlife sightings can be accurately georeferenced with the aid of high-accuracy positioning techniques. For example, smartphones equipped with high-accuracy GPS units ensure generated data record is associated with accurate geographic coordinates. Nevertheless, the above observations hold only for citizen observers who are reporting or recording data at the time of sighting wildlife occurrences in the field. In many cases, local residents (e.g., farmers) do not keep records of daily wildlife sightings or they simply do not have access to GPS units or smartphones. Most often, wildlife data are elicited from their memories long after the time of sighting [17, 18, 23, 40].

Wildlife data (e.g., sightings) collected from local citizens through interviews or questionnaire surveys often have position information with unsatisfactory accuracy [14, 15]. Descriptions of the locations of wildlife sightings are often imprecise or vague, particularly if a long time has lapsed since the actual sightings. Such incapability partly results from the absence of an effective interviewing media (e.g., an intuitive and interactive representation of the natural environment where local citizens are active) that facilitates local citizens to recall and locate their sightings of wildlife. Ref. [17] collected distribution and abundance data of terrestrial tortoises from local shepherds over 1 km × 1 km grid cells with the aid of topographic maps. However, it is difficult to accurately locate wildlife sightings on topographic maps for the local residents who had no training in map reading.

Geospatially enabled and user-friendly geovisualization interfaces could help improve positional accuracy of the wildlife data elicited from local residents [50, 51]. Geovisualization, particularly 3D geovisualization techniques, can be adopted to help local residents to recall and locate their sightings of wildlife and obtain wildlife data with more accurate positional information [40]. Given the flat 2D topographic maps, the local residents need relief interpretation skills to re-construct the 3D topography of the landscape; local residents can then orientate themselves and locate places on the landscape. But they often do not have much training in basic map reading skills, not to mention relief interpretation. 3D geovisualization can facilitate relief interpretation by producing a realistic and intuitive terrain representation [52] and improves visual search efficiency and navigation performance [53].

Geovisualization techniques as discussed above help improve positional accuracy of wildlife data at the very beginning of data generation. Sometimes, in cases where positional uncertainty does exist in wildlife data and is indeed of concern for wildlife habitat assessment, GIS-based methods have been developed to reduce its impact on the accuracy of wildlife habitat assessment. As an example, [54] proposed a spatial sampling method for deriving probable wildlife occurrence locations from patrol records using heuristics based on data recording context and species ecology to increase the accuracy of habitat suitability mapping.

3.2 Geospatial analysis to tackle spatial bias

Many geospatial analytical methods have been proposed to account for the spatial bias in wildlife data. An *AdaSTEM* approach that exploits variation in the density of wildlife observations was proposed to accommodate spatial bias in citizen-contributed wildlife observations [22, 55]. The continent- or hemisphere-wide study area is partitioned into rectangular spatial units (i.e., sub-areas) of size dependent upon density of wildlife observations. Environmental niches are modeled with only observations in each sub-area. By training local models in sub-areas, instead of training a global model using observations over the whole area, this approach mitigates spatial bias in the overall data set to a certain degree.

Filtering samples in the geographic or environmental space (i.e., remove observations that are within certain distance of one another) is also applied to reduce spatial bias [56, 57]. This method is based on the heuristic that removing localities (i.e., field samples) that are within certain distance of one another would somehow balance the unequal sampling or observation effort. The key of this method is to determine the distance threshold properly.

If detailed information observation effort is available, such information can then be incorporated to correct for spatial bias. Spatial bias in wildlife observations was compensated for by weighting the observations with weights inversely proportional to the cumulative visibility at the observation sites, given that cumulative visibility is a good proxy of the underlying observation effort [23]. Here, cumulative visibility is the frequency at which a given location can be seen by observers from the routes the observers take. It can be computed based on a digital elevation model (DEM) representing the terrain and the routes using viewshed analysis, a common GIS function.

A FactorBiasOut method was developed to correct for spatial bias in species presence-only data for species distribution modeling with MAXENT [58]. This method first estimates an empirical distribution to approximate the underlying but usually unknown sampling distribution that generated the presence-only data. This approximate sampling distribution is then used to factor out the spatial bias in presence-only data. This is achieved by feeding MAXENT with background data that have the same spatial bias as the presence data. For example, occurrence data of a target group of species that are observed by similar methods can be taken as the estimate of the effort information and thus are used as the background data.

Recently, a general representativeness-directed approach was proposed to spatial bias mitigation in citizen-contributed wildlife observations (i.e., samples) for habitat suitability mapping [59]. The key idea is to define and quantify the representativeness of samples and then properly weigh the samples to improve representativeness. Sample representativeness is defined as the "goodness-of-coverage" of the samples in the environmental covariate space, which in turn is measured by the similarity between the probability distribution of the samples in the covariate space and the probability distribution of all mapping units (e.g., pixels) within the study area. Spatial bias is then mitigated by weighting the samples toward increasing sample representativeness. The optimal weights that maximize sample representativeness are determined through an optimization procedure using a genetic algorithm.

3.3 Geocomputation to enable big data analysis

Citizen-contributed wildlife data are an important source of geospatial big data. In spatial analysis or modeling of such large volume of data (e.g., point pattern analysis, wildlife habitat assessment, and species distribution modeling), it is urgent to address the associated computational challenges. Geocomputation technologies could be utilized to address such computational challenges.

For example, over 100 million bird observations were added to the eBird database each year. Point pattern analysis is commonly used to discover patterns from such data. Existing point pattern analysis software tools are not able to handle geospatial big data efficiently. Cutting-edge geocomputation technologies such as cloud computing and GPU (graphics processing units) computing can be leveraged to accelerate point pattern analysis algorithms. The massively parallel computing powers of cloud computing and GPU computing effectively sped up point pattern analysis tasks on big data by a factor of hundreds [60, 61]. Given the significant acceleration brought by the geocomputation technologies, geospatial big data analysis tasks that once were computationally prohibitive can now be conducted in a timely manner.

4. Integrating citizen science and GIS for wildlife habitat assessment: a *Rhinopithecus bieti* case study

A case study of mapping black-and-white snub-nosed monkeys' (*Rhinopithecus bieti*) habitat suitability using *R. bieti* sighting data elicited from local villagers at Mt. Lasha in Yunnan, China, was presented to demonstrate the integration of citizen science and GIS for wildlife habitat assessment.

4.1 Species and study site

R. bieti is an "endangered" species on the IUCN (International Union for Conservation of Nature) Red List of Threatened Species [62]. *R. bieti* is endemic to the eastern Himalayas in northwest Yunnan and southeast Tibet, China, between the upper Mekong and Yangtze Rivers with 19 relatively isolated groups [63]. The monkeys use lichens (e.g., arboreal fruticose *Bryoria* and *Usnea* spp.) as their main food [64]. They prefer fir-larch forest at higher elevations in the

DOI: http://dx.doi.org/10.5772/intechopen.83681

northern part of the range but also stay in mixed coniferous and broad-leaf forest at lower elevations (above 2600 m) in the southern range [65]. Across its geographic distribution areas, the habitat of *R. bieti* has undergone degradation (e.g., habitat reduction and fragmentation) in the past decades due to the growth of human population that mostly employed traditional modes of production in its distribution area (e.g., clear-cutting forests for farming, grazing and firewood consumption; hunting) [65].

The study site is Mt. Lasha area located in northwest Yunnan Province, southwest China (**Figure 1**). Mt. Lasha is near the southern-most part of its geographic range [63, 65]. The 20.31 km^2 study area is an important habitat for a group of about 100 *R. bieti* individuals in 11 one-male multi-female units and two all-male units [66]. *R. bieti* is a significant species with a strong historic dimension in local communities [63]. On the one hand, hunting poses the greatest threat to the monkeys [63]. Local residents had long been hunting the monkeys for various purposes. Even after the Chinese government has designated the species in the first class of protected animals since 1977, illegal hunting had not been stopped completely. On the other hand, *R. bieti* habitat use is closely related to forest-provisioned resources including food and shelter [63, 65]. In the study area, two historical events had significant impacts on these forest-provisioned resources: in 1979 the China Environmental Protection Act was enacted; in 2006, the Mt. Lasha area became a protected area as part of the Yunling Nature Reserve. The forestry policy implementations associated with these events in the study area directly affected local residents' exploitation of the forests.

4.2 Data collection

4.2.1 Wildlife data elicited from local villagers

Sightings of *R. bieti* were elicited from local villagers for assessing the habitat of the monkeys in the study area across historical periods. Local villagers whose livelihoods are closely dependent on ecosystem services have long been living in the local area and accumulated information about *R. bieti* habitat use. Sightings of *R. bieti*

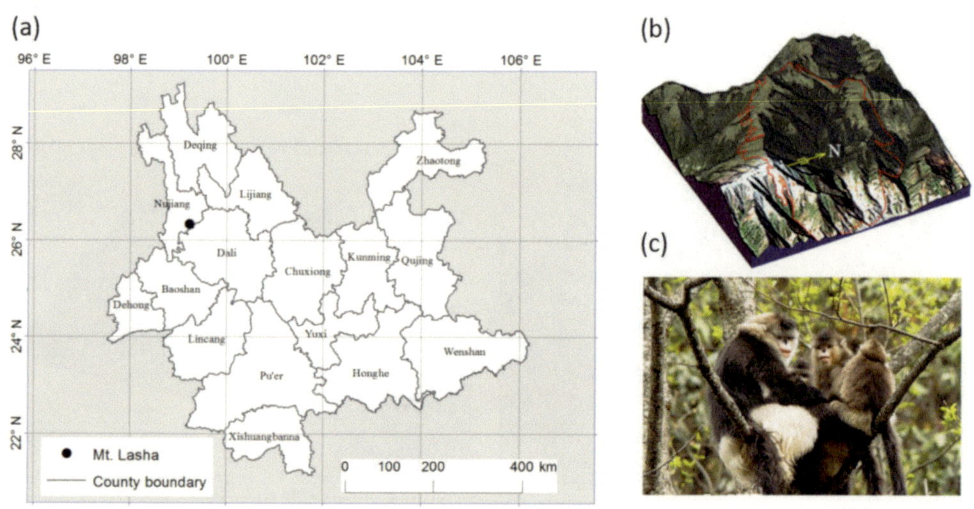

Figure 1.
Location of the study area: (a) Mt. Lasha in northwest Yunnan, China; (b) a 3D perspective image of the Mt. Lasha area; and (c) a family of R. bieti in their natural habitat (extracted from [40] with permission from John Wiley and Sons).

were elicited from local villagers through interviews. A 3D geovisualization tool was adopted to aid the interviews by using it to help the local villagers recall and locate where they sighted the monkeys more accurately.

R. bieti sightings were collected through structured interviews with local villagers (**Figure 2**). Sightings of the monkeys and activity routes of the villagers were elicited. The interviews were conducted using 3dMapper, a 3D geovisualization GIS tool that uses high-resolution DEM and satellite imagery to produce an intuitive 3D view of the study area [67] (freely available from solim.geography.wisc.edu). The user can zoom, pan, and easily draw points, lines, or polygons over the 3D scene. We introduced this geovisualization tool to the villagers to help them identify locations where they had sighted the monkeys and the daily routes they took in the area. The villagers also recalled the year and month when they sighted the monkeys or took the routes. The year of *R. bieti* sightings recalled by the interviewees was cross-checked with and refined with reference to timing of major events such as national policy implementations, date of marriage and child born, etc. and the month to seasonal activity patterns in the area such as farming and grazing. Information on where and when they sighted the monkeys was recorded as polygons. Information on the routes they took and the frequency with which they took each route was recorded as lines.

Geovisualization interview sessions were carried out by one biologist and one field assistant who were very familiar with the study area during July and August 2010. Sixty-eight local residents including herdsmen, hunters, and farmers who had extensive experience in the mountains from all five nearby villages were interviewed. The majority of them are aged between 30 and 60 (**Table 1**). The elicited *R. bieti* sightings and activity routes of the villagers cover a temporal span from the 1950s through 2010. Constrained by the availability of environmental data needed for habitat assessment, only *R. bieti* sightings in three historical periods (1973–1981, 1987–2005, and 2006–2010) were used for habitat assessment (see [40] for details) (**Figure 3**).

4.2.2 Environmental data

Environmental factors impacting *R. bieti* habitat use include terrain, water source, shelter and food, and human-posed disturbance [65, 68, 69]. Accordingly, the following environmental data layers were used in habitat assessment (habitat suitability mapping) for *R. bieti* in the study area [23, 40]: elevation, slope gradient, slope

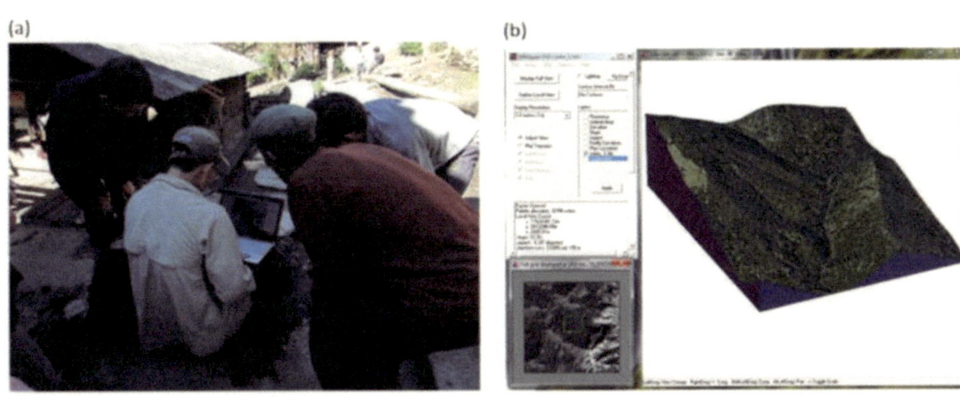

Figure 2.
Geovisualization interview sessions with the local residents using 3dMapper: (a) the local residents locating monkey sightings and activity routes and (b) a 3D scene of a small portion of the study area on which the local residents outlined monkey sightings and routes (extracted from [40] with permission from John Wiley and Sons).

Age	19–30	31–40	41–50	51–60	61–70	71–78
Count	7	12	16	18	10	5

Table 1.
Age composition of the interviewed local villagers.

aspect, distance to river, distance to village or road, and vegetation type. Interested readers can refer to [40] for details on how to obtain these environmental data layers.

4.3 Accounting for positional uncertainty and spatial bias

Data elicited from local villagers impose two challenges, namely positional uncertainty and spatial bias. First, local villagers often recall *R. bieti* sightings in the form of "I saw the monkeys over this area." Clearly, "over this area" can be depicted using a polygon, but this does not mean that the monkeys showed up at every location in the polygon area and certainly not at an equal probability within the polygon. Thus, taking all locations in the polygon as sightings is not appropriate. The question then is how to obtain locations that are representative of the actual presence of wildlife in the area outlined by the local villagers. The second challenge is the spatial bias in the elicited *R. bieti* sightings due to local villagers' opportunistic observation effort. For example, multiple sightings of monkeys at one location by many villagers do not necessarily mean that the location is highly preferred by the monkeys; it might be that the location is easily visible from multiple activity routes. Thus, every time a monkey shows up at this location, it is spotted by some villager(s). On the other hand, a monkey that shows up at locations that are preferred by monkeys but less visible to the villagers will have a lesser chance of being spotted. This spatial bias must be compensated for when using the elicited *R. bieti* sightings for wildlife habitat assessment.

Geospatial analysis methods provisioned by GIS were adopted to address the two challenges. First, a frequency sampling strategy [23, 70] was applied to reduce the

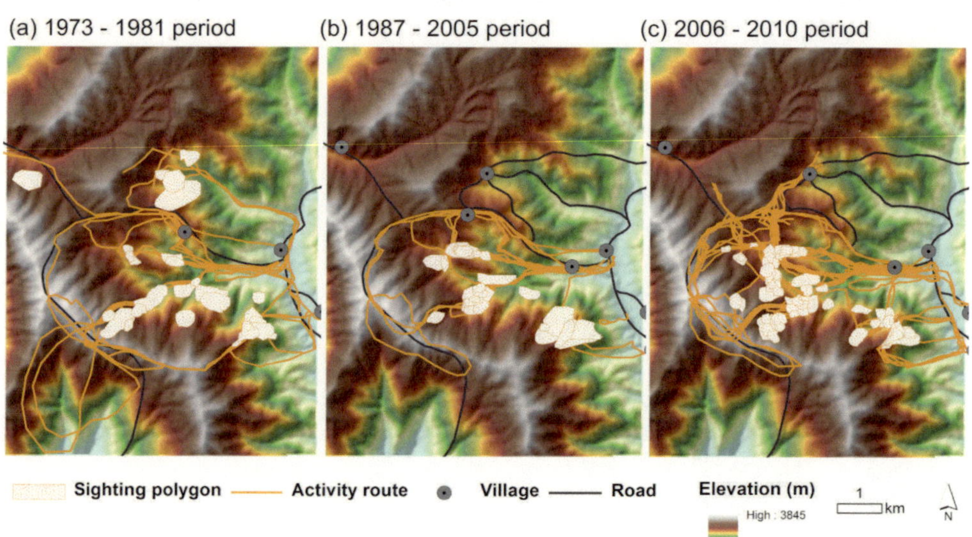

Figure 3.
Sightings of R. bieti and activity routes elicited from the local residents through geovisualization interviews: (a) 1973–1981 period; (b) 1987–2005 period; (c) 2006–2010 period (extracted from [40] with permission from John Wiley and Sons).

position uncertainty in sighting polygons provided by local villagers and to identify the representative locations for *R. bieti* presence within each polygon. It is assumed that locations at which values of environmental conditions are most frequent over the polygon area would approximate the locations of actual presence best. Under this assumption, the frequency sampling strategy implemented in GIS was applied to locate the representative locations within a polygon. Here, only the general idea was outlined as above; full details on implementing the sampling strategy in GIS can be found in [23, 70].

Second, the spatial bias was compensated for by inversely weighting each representative presence location with cumulative visibility of the location from the routes taken by local villagers [23]. In this particular case study, spatial bias in the elicited *R. bieti* sightings was a result of the non-random and uneven distribution of local villagers' observation efforts constrained by activity routes and terrain (as discussed in depth in Section 2.3.3). Thus, cumulative visibility was treated as a proxy of the underlying observation effort and can be incorporated to compensate for spatial bias. The cumulative visibility of a location can be computed in GIS based on a DEM and the activity routes of local villagers [23]. The efficacy of the frequency sampling strategy to reduce positional uncertainty and the visibility-weighting scheme to compensate for spatial has been demonstrated in [23].

4.4 Habitat assessment

A kernel density estimation-based method [23, 71] was applied to derive the relationship between *R. bieti* habitat suitability and environmental conditions. This method estimates a probability density function representing the probability distribution of wildlife presence over the gradient of each environmental factor based on the values of the environmental factors over the presence locations. In estimating the probability density functions, presence locations are weighted by the in situ cumulative visibility from activity routes of the local villagers to compensate for spatial bias. The probability density functions are then normalized to the range of [0, 1] to represent the relationships between habitat suitability and individual environmental factors (**Figure 4**). The overall habitat suitability considering all

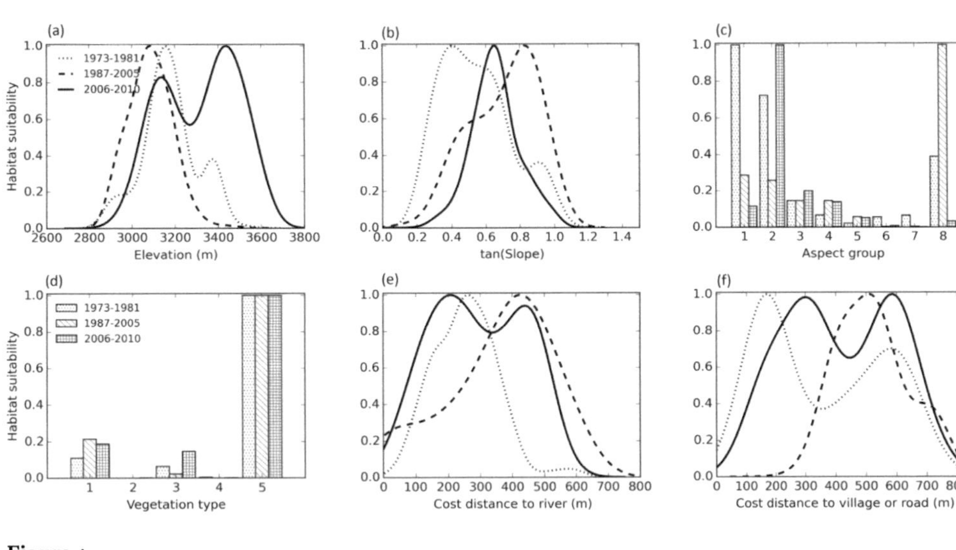

Figure 4.
Suitability-environment relationships derived from elicited R. bieti sightings in each historical period. Aspect group 1: 0–45° (starting from north), 2: 45–90°, 3: 90–135°, 4: 135–180°, 5: 180–225°, 6: 225–270°, 7: 270–315°, 8: 315–360°. Vegetation type 1: evergreen coniferous, 2: pasture, 3: yunnan pine, 4: farmland, 5: deciduous broadleaf (extracted from [40] with permission from John Wiley and Sons).

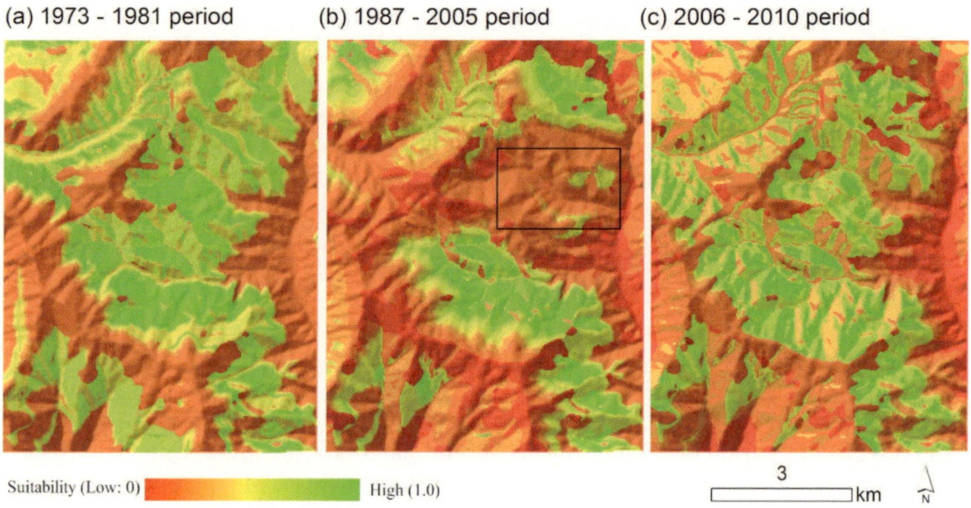

Figure 5.
Habitat suitability maps predicted for the study area using elicited R. bieti sightings in each historical period (a) 1973–1981 period; (b) 1987–2005 period; (c) 2006–2010 period (extracted from [40] with permission from John Wiley and Sons).

environmental factors is determined by integrating the relationships based on a "limiting factor" principle (see [23] for full details of the method). Computing the overall habitat suitability at every location (pixel) in the study area resulted in habitat suitability maps as shown in **Figure 5**.

Across the three historical periods, high suitability habitats were in forests (**Figure 4d**) at mid-to-high elevation range (**Figure 4a**) on the northeast hill slopes (**Figure 4c**). Overall, high suitability habitats shrank in the 1987–2005 period compared to the previous period. As an example, the area outlined on **Figure 5b** in the 1987–2005 period is of much lower suitability compared to the 1973–1981 period. This might be a result of the introduction of new village settlements and roads in that area in the 1987–2005 period which induced significant human disturbance. *R. bieti* habitats were recovering in the 2005–2010 period. The outlined area recovered to higher suitability in that period; this might be attributed to the monkeys getting used to proximity to villages and roads (**Figure 4f**).

The derived relationships between *R. bieti* habitat suitability and individual environmental factors (**Figure 4**) confirmed the recovering trend in the 2006–2010 period. The elevation range of high suitability habitats in the 2006–2010 period shifted back to higher ranges close to those in the 1973–1981 period (**Figure 4a**). The ranges of distance to rivers and distance to village or road corresponding to high suitability habitats also shifted back to similar ranges as in the 1973–1981 period (**Figure 4e, f**). These were potential evidences that conservation practices initiated by the Yunling Nature Reserve have restored *R. bieti* habitat in the area.

5. Conclusions

Wildlife data required for wildlife habitat assessment can be difficult and expensive to obtain with traditional data collection methods (e.g., biological survey, geographic sampling), especially for conservation programs with limited budget support in remote and poor areas. Citizen science offers a cost-effective way of collecting wildlife data to sustain such programs. Nevertheless, average citizens are non-professionals and their wildlife observation efforts are un-coordinated. Thus, wildlife data contributed by citizens may be subject to data quality issues such as positional uncertainty and spatial bias. This chapter provides an overview of citizen science as a means of collecting wildlife data, GIS-provisioned geovisualization, and geospatial analysis techniques for tackling the data quality issues of citizen-contributed wildlife data, and the integration of citizen science and GIS for wildlife habitat assessment. A case study of mapping *R. bieti* habitat suitability using *R. bieti* sightings elicited from local villagers in Yunnan, China, was presented as an example to demonstrate the usefulness of integrating citizen science and GIS for wildlife habitat assessment.

Acknowledgements

Support to the author from the University of Denver through the Faculty Startup Funds is greatly appreciated. The author would like to thank Dr. A-Xing Zhu in the Department of Geography at the University of Wisconsin-Madison for his invaluable inputs on designing methods, and Dr. Wen Xiao and Dr. Zhi-Pang Huang in the Institute of Eastern-Himalaya Biodiversity Research at Dali University and officials in the Yunling Nature Reserve administration for their generous support for field work of the reported *R. bieti* case study.

Conflict of interest

The author declares no conflict of interest.

Author details

Guiming Zhang
Department of Geography and the Environment, University of Denver, Denver, United States

*Address all correspondence to: guiming.zhang@du.edu

References

[1] Reunanen P, Monkkonen M, Nikula A. Habitat requirements of the Siberian flying squirrel in northern Finland: Comparing field survey and remote sensing data. Annales Zoologici Fennici. 2002;**39**(1):7-20

[2] Pulliam HR, Danielson BJ. Sources, sinks, and habitat selection: A landscape perspective on population dynamics. The American Naturalist. 1991;**137**(s1):S50-S66

[3] Araújo MB, Williams PH. Selecting areas for species persistence using occurrence data. Biological Conservation. 2000;**96**(3):331-345

[4] Guisan A, Tingley R, Baumgartner JB, Naujokaitis-Lewis I, Sutcliffe PR, Tulloch AIT, et al. Predicting species distributions for conservation decisions. Ecology Letters. 2013;**16**(12):1424-1435

[5] Thorn JS, Nijman V, Smith D, Nekaris K a I. Ecological niche modelling as a technique for assessing threats and setting conservation priorities for Asian slow lorises (primates: Nycticebus). Diversity and Distributions. 2009;**15**(2):289-298

[6] Thuiller W, Richardson DM, Pyšek P, Midgley GUYF, Hughes GO, Rouget M. Niche-based modelling as a tool for predicting the risk of alien plant invasions at a global scale. Global Change Biology. 2005;**11**(12):2234-2250

[7] Franklin J, Miller JA. Mapping Species Distributions: Spatial Inference and Prediction. Vol. 338. Cambridge: Cambridge University Press; 2009

[8] Rotenberry JT, Preston KL, Knick ST. GIS-based niche modeling for mapping species' habitat. Ecology. 2006;**87**(6):1458-1464

[9] Viña A, Bearer S, Zhang H, Ouyang Z, Liu J. Evaluating MODIS data for mapping wildlife habitat distribution. Remote Sensing of Environment. 2008;**112**(5):2160-2169

[10] Hijmans RJ, Cameron SE, Parra JL, Jones PG, Jarvis A. Very high resolution interpolated climate surfaces for global land areas. International Journal of Climatology. 2005;**25**(15):1965-1978

[11] Campbell AF, Sussman RW. The value of radio tracking in the study of neotropical rain forest monkeys. American Journal of Primatology. 1994;**32**(4):291-301

[12] Trolle M, Kéry M. Estimation of ocelot density in the pantanal using capture-recapture analysis of camera-trapping data. Journal of Mammalogy. 2003;**84**(2):607-614

[13] Hulbert IAR, French J. The accuracy of GPS for wildlife telemetry and habitat mapping. Journal of Applied Ecology. 2001;**38**(4):869-878

[14] Msoffe F, Mturi F, Galanti V, Tosi W, Wauters L, Tosi G. Comparing data of different survey methods for sustainable wildlife management in hunting areas: The case of Tarangire–Manyara ecosystem, northern Tanzania. European Journal of Wildlife Research. 2007;**53**(2):112-124

[15] Xu F, Ma M, Wu YQ, Chundawat RS. Distribution of the ibex (Capra ibex) in Tomur National Nature Reserve of Xinjiang, China. Zoological Research. 2007;**28**(6):670-672

[16] Myers N, Mittermeier RA, Mittermeier CG, da Fonseca GAB, Kent J. Biodiversity hotspots for conservation priorities. Nature. 2000;**403**(6772):853-858

[17] Anadón JD, Giménez A, Ballestar R, Pérez I. Evaluation of local ecological knowledge as a method for collecting extensive data on animal abundance. Conservation Biology. 2009;**23**(3):617-625

[18] Anadón JD, Giménez A, Ballestar R. Linking local ecological knowledge and habitat modelling to predict absolute species abundance on large scales. Biodiversity and Conservation. 2010;**19**(5):1443-1454

[19] Dickinson JL, Shirk J, Bonter D, Bonney R, Crain RL, Martin J, et al. The current state of citizen science as a tool for ecological research and public engagement. Frontiers in Ecology and the Environment. 2012;**10**(6):291-297

[20] Goodchild MF. Citizens as sensors: The world of volunteered geography. GeoJournal. 2007;**69**(4):211-221

[21] Sullivan BL, Wood CL, Iliff MJ, Bonney RE, Fink D, Kelling S. eBird: A citizen-based bird observation network in the biological sciences. Biological Conservation. 2009;**142**(10):2282-2292

[22] Fink D, Damoulas T, Dave J, Damoulas T, Dave J. Adaptive spatio-temporal exploratory models: Hemisphere-wide species distributions from massively crowdsourced eBird data. In: Twenty-Seventh AAAI Conference on Artificial Intelligence (AAAI-13). 2013. pp. 1284-1290

[23] Zhu A-X, Zhang G, Wang W, Xiao W, Huang Z-P, Dunzhu G-S, et al. A citizen data-based approach to predictive mapping of spatial variation of natural phenomena. International Journal of Geographical Information Science. 2015;**29**(10):1864-1886

[24] Eitzel MV, Cappadonna JL, Santos-Lang C, Duerr RE, Virapongse A, West SE, et al. Citizen science terminology matters: Exploring key terms. Citizen Science: Theory and Practice. 2017;**2**(1):1

[25] OED. Citizen science [Internet]. Oxford English Dictionary. 2018. Available from: http://www.oed.com/view/Entry/33513?redirectedFrom=citizen+science#eid316619123 [Accessed: November 14, 2018]

[26] Silvertown J. A new dawn for citizen science. Trends in Ecology & Evolution. 2009;**24**(9):467-471

[27] Follett R, Strezov V. An analysis of citizen science based research: Usage and publication patterns. PLoS One. 2015;**10**(11):1-14

[28] Butcher GS, Fuller MR, McAllister LS, Geissler PH. An evaluation of the Christmas bird count for monitoring population trends of selected species. Wildlife Society Bulletin. 1990;**18**(2):129-134

[29] Sauer JR, Hines JE, Fallon JE, Link WA, Fallon JE, Pardieck KL, et al. The North American breeding bird survey 1966-2011: Summary analysis and species accounts. North American Fauna. 2013;**79**(79):1-32

[30] Snäll T, Kindvall O, Nilsson J, Pärt T. Evaluating citizen-based presence data for bird monitoring. Biological Conservation. 2011;**144**(2):804-810

[31] Sullivan BL, Phillips T, Dayer AA, Wood CL, Farnsworth A, Iliff MJ, et al. Using open access observational data for conservation action: A case study for birds. Biological Conservation. 2017;**208**:5-14

[32] Catlin-Groves CL. The citizen science landscape: From volunteers to citizen sensors and beyond. International Journal of Zoology. 2012;**2012**:1-14

[33] Newman G, Wiggins A, Crall A, Graham E, Newman S, Crowston K. The future of citizen science: Emerging technologies and shifting paradigms. Frontiers in Ecology and the Environment. 2012;**10**(6):298-304

[34] Haklay M. Citizen science and volunteered geographic information: Overview and typology of participation. In: Sui D, Elwood S, Goodchild M, editors. Crowdsourcing Geographic Knowledge: Volunteered Geographic Information (VGI) in Theory and Practice. Dordrecht: SpringerNetherlands; 2013. pp. 105-122

[35] Wikipedia Contributors. eBird [Internet]. Wikipedia, The Free Encyclopedia. 2018. Available from: https://en.wikipedia.org/w/index.php?title=EBird&oldid=856114315 [Accessed: November 25, 2018]

[36] Hochachka WM, Fink D, Hutchinson RA, Sheldon D, Wong W-K, Kelling S. Data-intensive science applied to broad-scale citizen science. Trends in Ecology & Evolution. 2012;**27**(2):130-137

[37] Kelling S, Hochachka WM, Fink D, Riedewald M, Caruana R, Ballard G, et al. Data-intensive science: A new paradigm for biodiversity studies. Bioscience. 2009;**59**(7):613-620

[38] Miller HJ, Goodchild MF. Data-driven geography. GeoJournal. 2014;**80**(4):449-461

[39] Kelling S, Lagoze C, Wong W-K, Yu J, Damoulas T, Gerbracht J, et al. eBird: A human/computer learning network to improve biodiversity conservation and research. AI Magazine. 2013;**34**(1):10-20

[40] Zhang G, Zhu A-X, Huang Z-P, Ren G, Qin C-Z, Xiao W. Validity of historical volunteered geographic information: Evaluating citizen data for mapping historical geographic phenomena. Transactions in GIS. 2018;**22**(1):149-164

[41] Brown G, McAlpine C, Rhodes J, Lunney D, Goldingay R, Fielding K, et al. Assessing the validity of crowdsourced wildlife observations for conservation using public participatory mapping methods. Biological Conservation. 2018;**227**(September):141-151

[42] Ellwood ER, Bart HL, Doosey MH, Jue DK, Mann JG, Nelson G, et al. Mapping life—Quality assessment of novice vs. expert georeferencers. Citizen Science: Theory and Practice. 2016;**1**(1):1-12

[43] Sauer JR, Peterjohn BG, Link WA. Observer differences in the North American breeding bird survey. Auk. 1994;**111**(1):50-62

[44] Kendall WL, Peterjohn BG, Sauer JR, Url S. First-time observer effects in the North American breeding bird survey. Auk. 1996;**113**(4):823-829

[45] Freitag A, Meyer R, Whiteman L. Strategies employed by citizen science programs to increase the credibility of their data. Citizen Science: Theory and Practice. 2016;**1**(2):1-11

[46] Osborne PE, Leitão PJ. Effects of species and habitat positional errors on the performance and interpretation of species distribution models. Diversity and Distributions. 2009;**15**(4):671-681

[47] Kadmon R, Farber O, Danin A. Effect of roadside bias on the accuracy of predictive maps produced by bioclimatic models. Ecological Applications. 2004;**14**(2):401-413

[48] Boakes EH, McGowan PJK, Fuller RA, Ding C, Clark NE, O'Connor K, et al. Distorted views of biodiversity: Spatial and temporal bias in species occurrence data. PLoS Biology. 2010;**8**(6):e1000385

[49] Leitão PJ, Moreira F, Osborne PE. Effects of geographical data sampling bias on habitat models of species distributions: A case study with steppe birds in southern Portugal. International

Journal of Geographical Information Science. 2011;**25**(3):439-454

[50] Seeger CJ. The role of facilitated volunteered geographic information in the landscape planning and site design process. GeoJournal. 2008;**72**(3-4):199-213

[51] Newman G, Zimmerman D, Crall A, Laituri M, Graham J, Stapel L. User-friendly web mapping: Lessons from a citizen science website. International Journal of Geographical Information Science. 2010;**24**(12):1851-1869

[52] Carbonell Carrera C, Bermejo Asensio LA. Augmented reality as a digital teaching environment to develop spatial thinking. Cartography and Geographic Information Science. 2017;**44**(3):259-270

[53] Liao H, Dong W, Peng C, Liu H. Exploring differences of visual attention in pedestrian navigation when using 2D maps and 3D geo-browsers. Cartography and Geographic Information Science. 2017;**44**(6):474-490

[54] Zhang G, Zhu A-X, Huang Z-P, Xiao W. A heuristic-based approach to mitigating positional errors in patrol data for species distribution modeling. Transactions in GIS. 2018;**22**(1):202-216

[55] Fink D, Hochachka WM, Zuckerberg B, Winkler DW, Shaby B, Munson MA, et al. Spatiotemporal exploratory models for broad-scale survey data. Ecological Applications. 2010;**20**(8):2131-2147

[56] Boria RA, Olson LE, Goodman SM, Anderson RP. Spatial filtering to reduce sampling bias can improve the performance of ecological niche models. Ecological Modelling. 2014;**275**:73-77

[57] Varela S, Anderson RP, García-Valdés R, Fernández-González F. Environmental filters reduce the effects of sampling bias and improve predictions of ecological niche models. Ecography. 2014;**37**(11):1084-1091

[58] Dudík M, Phillips SJ, Schapire RE, Dudik M, Schapire RE, Phillips SJ, et al. Correcting sample selection bias in maximum entropy density estimation. Advances in neural information processing systems. 2005;**17**:323-330

[59] Zhang G. A Representativeness Directed Approach to Spatial Bias Mitigation in VGI for Predictive Mapping [Thesis]. Madison: University of Wisconsin-Madison; 2018

[60] Zhang G, Huang Q, Zhu A-X, Keel J. Enabling point pattern analysis on spatial big data using cloud computing: Optimizing and accelerating Ripley's K function. International Journal of Geographical Information Science. 2016;**30**(11):2230-2252

[61] Zhang G, Zhu A-X, Huang Q. A GPU-accelerated adaptive kernel density estimation approach for efficient point pattern analysis on spatial big data. International Journal of Geographical Information Science. 2017;**31**(10):2068-2097

[62] IUCN. The IUCN red list of threatened species version 2018-2 [Internet]. 2018. Available from: https://www.iucnredlist.org [Accessed: December 05, 2018]

[63] Long YC, Kirkpatrick CR, Zhong T, Xiao L. Report on the distribution, population, and ecology of the Yunnan snub-nosed monkey (*Rhinopithecus bieti*). Primates. 1994;**35**(2):241-250

[64] Kirkpatrick RC, Long YC, Zhong T, Xiao L. Social organization and range use in the Yunnan snub-nosed monkey *Rhinopithecus bieti*. International Journal of Primatology. 1998;**19**(1):13-51

[65] Xiao W, Ding W, Cui LW, Zhou RL, Zhao QK. Habitat degradation of *Rhinopithecus bieti* in Yunnan, China.

International Journal of Primatology. 2003;**24**(2):389-398

[66] Huang ZP, Cui LW, Scott M, Wang SJ, Xiao W. Seasonality of reproduction of wild black-and-white snub-nosed monkeys (*Rhinopithecus bieti*) at Mt. Lasha, Yunnan, China. Primates. 2012;**53**(3):237-245

[67] Burt JE, Zhu AX. 3dMapper 4.02. 4.02. In: Terrain Analytics. Madison, WI: LLC; 2004

[68] Long Y, Zhong T, Xiao L. Study on geographical distribution and population of the Yunnan snub-nosed monkey. Zoological Research. 1996;**17**(4):437-441

[69] Huang ZP. Foraging, reproduction and sleeping site selection of black-and-white snub-nosed monkey (*Rhinopithecus bieti*) at the southern range. [master's dissertation]. Kunming: Faculty of Conservation Biology, Southwest Forestry University; 2009

[70] Qi F, Zhu A-X. Knowledge discovery from soil maps using inductive learning. International Journal of Geographical Information Science. 2003;**17**(8):771-795

[71] Zhang G, Zhu A-X, Windels SK, Qin C-Z. Modelling species habitat suitability from presence-only data using kernel density estimation. Ecological Indicators. 2018;**93**:387-396

Chapter 3

Ecology of Feral Pigeons: Population Monitoring, Resource Selection, and Management Practices

Erin E. Stukenholtz, Tirhas A. Hailu, Sean Childers, Charles Leatherwood, Lonnie Evans, Don Roulain, Dale Townsley, Marty Treider, R. Neal Platt II, David A. Ray, John C. Zak and Richard D. Stevens

Abstract

Feral pigeons (*Columba livia*) are typically ignored by ornithologists but can be found roosting in the thousands within cities across the world. Pigeons have been known to spread zoonoses, through ectoparasites and excrement they produce. Along with disease, feral pigeons have an economic impact due to the cost of cleanup and maintenance of human infrastructure. Many organizations have tried to decrease pigeon abundances through euthanasia or use of chemicals that decrease reproductive output. However, killing pigeons has been unsuccessful in decreasing abundance, and chemical inhibition can be expensive and must be used throughout the year. A case study at Texas Tech University has found that populations fluctuate throughout the year, making it difficult to manage numbers. To successfully decrease populations, it is important to have a multifaceted approach that includes removing necessary resources (i. e. nest sites and roosting areas) and decreasing the number of offspring through humane techniques.

Keywords: birth control, nest sites, nuisance, rock doves, zoonoses

1. Introduction

Of the 7.53 billion people that live on Earth, over half inhabit cities [1, 2]. Increase in development has altered biodiversity through an increase in fragmentation and invasive species abundance. Urban areas are highly susceptible to invasions of nonnative species [3], which can increase threat to native species and increase economic costs due to environmental and structural damage [4, 5]. One species that is associated with urban areas, but is often ignored by ornithologists, is the invasive feral pigeon (*Columba livia*) [6].

Descendants of cliff-dwelling rock pigeons, feral pigeons have a close relationship with humans [7]. Most likely originating in southern Asia, wild rock pigeons colonized areas in Europe and North Africa [7]. Around 5,000–10,000 years ago, rock pigeons became the first known domesticated bird [8]. Throughout history, these domesticated birds were honored for their reproductive ability and because they were an important food source [9]. As agriculture developed, the use of pigeons for meat dwindled, leading to the species escaping and creating feral flocks in urban areas across the world [9].

Humans did not consider pigeons to be a nuisance until the twentieth century [10]. Pigeon abundance is positively correlated with human density [11]. Because of the increase in human abundance and activity, pigeon ability to exploit many different food types [5, 12] and this combined with low predation risks caused an increase in pigeon population growth [13]. Buildings found in cities closely resemble cliffs and pose a synthetic habitat for pigeons [14]. Moreover, cities provide a wide variety of resources for pigeons, from nesting [15] and roosting locations [5] to increased ambient temperature [11]. Nowadays, they can be seen roosting in the thousands throughout many cities [10].

1.1 Pigeon problem

With an increase in abundance comes an increase in economic impacts and health issues associated with pigeons. Each year in the United States, pigeons cause approximately 1.1 billion dollars in environmental and infrastructural damage [4]. For example, pigeons residing in cities that are surrounded by agriculture often steal grain from nearby silos [6]. One such warehouse lost 3 tons of grain a week due to pigeon theft [6]. Feral pigeons can also cause building damage. An individual pigeon can excrete up to 12 kg of excrement a year [16]. Due to their human-modified diet, feral pigeons typically have more acidic excreta compared to wild rock pigeons [17]. The compounding effects of high pigeon abundance, large amounts of excreta, and high levels of acidity can have devastating results for building structures. The acidity alone can cause structural and esthetic damage over time [7]. Feces and nest material can clog drainage systems, causing internal damage to buildings [7]. The costs to clean and repair structural damage caused by pigeons can be straining on family and commercial businesses [7].

1.2 Human impacts

Damage to buildings is insignificant when it comes to health concerns that pigeons pose. Many citizens view pigeons as pests that can spread plague-like diseases via excreta, secretions, parasites, and dust from their feathers [7, 18]. Diseases can be spread by excrement and dust, and thus direct contact is not needed for transmission [19]. Diseases that can be potentially transmitted from pigeons to humans include aspergillosis, borreliosis, coccidiosis, chlamydiosis, equine encephalitis, influenza, paramyxovirus, paratyphoid, toxoplasmosis, and tuberculosis [19, 20], and some can potentially be lethal [18]. How often these diseases spread to humans is uncertain [7]. The most common pathogen transmitted to humans via pigeons is *Chlamydophila psittaci* [21]. The fungus *Histoplasma capsulatum* found in pigeon excrement is even more concerning [7]. With a large amount of excreta present in areas of high human activity, the possibility of humans becoming infected is high [7]. Maintenance workers have been infected with *histoplasmosis* after cleaning up pigeon excrement [22, 23].

1.3 Impacts on other species

Pigeons are a vector of several diseases that can be a hazard to domestic fowl and native species within cities. The paramyxovirus is a highly contagious respiratory

DOI: http://dx.doi.org/10.5772/intechopen.84612

infection that is fatal among birds [24]. In the 1980s, a large outbreak of paramyxovirus started in Italy and spread throughout Europe [25]. The outbreak in Great Britain in 1984 had devastating economic costs due to pigeons transmitting the disease to chickens [27]. In 1985, approximately 30% of pigeons were infected with the disease in Germany [26]. Nowadays, domestic fowl and other species of birds are still at risk from paramyxovirus outbreaks [28].

In invasive Columbiformes (avian order of feral pigeons), disease transmission tends to have the highest impact on other avian species [23, 29]. Research on how pigeons interact with other species via competition is limited. A study on direct competition among food patches of native and invasive birds in South Wales found that feral pigeons were the only exotic species that could potentially be considered an aggressive competitor [30].

There is no denying that invasive feral pigeons impact the lives of humans, through food theft, building damage, and increased risk of disease transmissions. Due to adverse effects of pigeons, businesses and campuses have tried to decrease their population size.

2. Case study of Texas Tech campus

Reducing pigeon numbers requires a multifaceted approach based on robust ecological information. Texas Tech University (TTU) located in Lubbock, Texas has had a history with pigeons. In the 1950s, TTU would host annual pigeon hunts on campus to decrease pigeon abundance and to supply food for the Christmas celebration at Milan's Children's home [31]. In the winter of 1957, 475 pigeons were shot on campus for the annual hunt [31]. Today, the main campus has approximately 100 Spanish Renaissance adorned buildings [32] on a 744 ha lot [33]. According to the physical plant at TTU, around $200,000 is spent each year cleaning up after pigeons on campus (Sean Childers, personal communication).

Because of the adverse effects caused by pigeons, TTU is conducting a detailed ecological study of their population residing on campus. The purpose of this study is to obtain estimates of population size. This will form a baseline to determine if there are significant decreases in abundance due to management strategies. Secondly, we will be able to recognize areas that have a high abundance of pigeons (hot spots), where we can then impose population reduction methods more aggressively. Thirdly, we will study how populations fluctuate throughout the year due to seasonal changes in the environment.

In 2017, we estimated abundance to be approximately 9819–13,757 pigeons [34]. We also found hot spots scattered across campus that had 649–750 pigeons [34]. To reduce the population, TTU is planning a management strategy that includes habitat modification and sterilization techniques. The estimated abundance can be used to determine if the strategy is effective and if there is a significant reduction over time. Hot spots in abundance can assist in determining where habitat modification and sterilization processes can be most effective in decreasing population size.

Along with enumerating pigeon abundance, we characterized population variation over time (**Figure 1**). By examining two buildings, the Experimental Sciences Building (ESB) and Holden Hall (HH), on campus that housed large pigeon populations, we found that there were daily and seasonal differences in numbers [35]. Abundance typically increased in the winter, except for ESB in the morning. Normally, morning counts for ESB were higher than afternoon counts [35]. Holden Hall pigeon abundance fluctuated from morning to the afternoon but increased throughout the year [35].

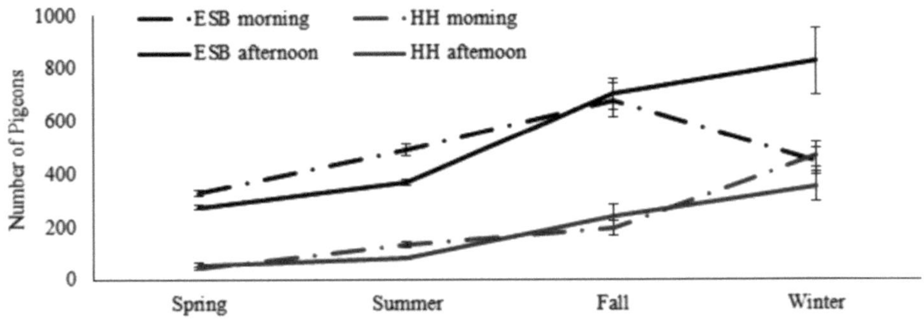

Figure 1.
Seasonal counts in the morning and afternoon of pigeons on two buildings on Texas Tech campus in 2017. ESB = experimental science building, HH = holden hall. Standard error ± 1.

Information on fluctuations in pigeon abundance can help when trying to organize a successful management plan. For example, we can determine if decreases in pigeon abundance are due to seasonal variation or due to management strategies, or when efforts to decrease abundance could be eased or made more active. However, to determine the most effective management for decreasing pigeon numbers, further understanding of ecology is necessary.

3. How to relieve the pigeon problem?

3.1 Removal of pigeons

One tactic to rid pigeons of an area is by humanely capturing them and transferring them to a new location [7]. This strategy has been proven to be ineffective, due to biological characteristics of pigeons. Pigeons have an excellent homing ability. Immediately after fledging, pigeons become imprinted to the site [36]. When taken to an unknown location, a pigeon will orient itself by using the Sun's orientation [37], Earth's geomagnetism [38], environmental odors [39], and landmarks [40] to navigate themselves homeward.

3.2 Killing pigeons

Another tactic to rid pigeons of an area is through euthanasia or culling. However, the tactic often is ineffective at decreasing pigeon abundance [18]. In Basel, Switzerland in the 1980s, pigeon populations were reduced by 80% through euthanasia. Within a few weeks, pigeon populations returned to the original size or even larger than before [41]. The removal of individuals opened up space for new migrants to come in, such as juveniles [41]. Because in urban areas pigeons reproduce throughout the year and have reduced mortality rates due to reduced predation, flock sizes can exceed carrying capacities and competition for breeding sites can increase [42].

In Stuttgart, Arkansas, controlled hunting was used to decrease pigeon abundance. Following the hunt, pigeon numbers decreased dramatically but within a few months rose to original levels [43]. From 1983 to 1991, Barcelona, Spain started a program of capturing and euthanizing pigeons. One single capture caught 362 individuals. However, pre- and post-census showed no significant difference in abundance [44]. Killing pigeons has proven to be only a temporary solution. Due to their high fertility rates, pigeons can rebound quickly.

3.3 Modifying habitat

A management plan can become successful by carefully studying what an organism selects for within its habitat. Pigeons are granivorous [7] and need a large feeding source to support their large population and offspring [7]. Furthermore, if a pigeon does not eat in 3 days, it can lose up to 5% of its body weight [45].

When killing pigeons proved ineffective for the city of Basel, the local government tried to decrease abundance by decreasing pigeon food supply [41]. In their situation, people created a huge food base for pigeons. They instituted an outreach program to educate people on the harm of feeding pigeons and the importance of maintaining a small population [41]. The decrease in food supply increased competition, decreased the probability of juveniles entering the population, and decreased reproductive success due to unfavorable conditions [41]. Within 4 years, one of the central flock sizes went from 1400 to 708 pigeons [41].

However, finding the primary source of food for pigeons may be difficult. Pigeons are opportunistic species [7] and have the ability to exploit a variety of food types [12], even human refuge [7]. Therefore, pigeons will feed within cities, consuming food left behind by humans, and can also travel to the outskirts of urban areas to eat from agricultural fields [7]. While it is common to see pigeons eating food distributed by humans, this rarely represents a high percentage of their diet [7]. Instead, pigeons are more likely to select grain from agricultural sources or from wild plants [7]. Therefore, simply limiting the food source may only have a limited effect. Moreover, such an effect depends on the foraging behavior of the flock [7]. If the majority of a flock feeds on a source that can potentially be managed, then it is possible to decrease their abundance [7].

If an organization has a population that forages outside their property, similar to the situation at TTU, influencing nest sites may be a more favorable way to decrease abundance than limiting the food source. As we have mentioned in this chapter, pigeons are prolific breeders. For most avian species, the breeding period is seasonal [46]. Pigeons have the advantage of being granivorous, where their food supply can survive during the winter months [7]. This consistent food supply helps pigeons feed their offspring with crop milk throughout the year [7]. Interestingly, wild rock pigeons have a seasonal breeding period, while feral pigeons breed throughout the year, even in winter (**Figure 2**) [46]. The ability to do so is probably due to the selective breeding done by humans in the past to produce a bird that can breed throughout the year [47].

The typical clutch size of feral pigeons is two eggs, very rarely is it more or less [7]. The incubation period lasts approximately 2 weeks, and depending on region and season, chicks will fledge from approximately 25–40 days after hatching [6, 7].

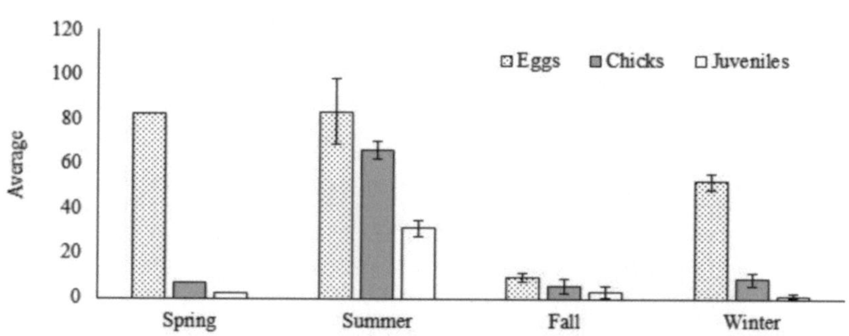

Figure 2.
Average number of eggs, chicks and juvenile pigeons on the experimental sciences building at Texas Tech University in 2017.

In Manchester, eggs hatched were approximately 62%, and chicks that fledged were around 66% [6]. Once chicks reached about 20 days of age, pigeons started laying another clutch [6]. With high reproductive success and year-long output, pigeon abundance can quickly increase.

Managing pigeon habitat via nest site removal could decrease reproductive output and decrease abundance. First, it is advantageous to locate areas that are selected for in terms of nesting locations. Pigeons typically nest in cavity-like structures on top of or within buildings (**Figure 3**) [7]. Any site that is covered and semi-dark provides a favorable site for pigeons to nest [7]. Nests can be found from inside attics to the underside of bridges and overpasses [7]. On TTU campus, pigeon nests can be found under heating, ventilation, and air conditioning (HVAC) equipment on top of buildings (**Figure 3**) that are protected from the elements and can provide a stable environment (e.g. temperature).

Modifying nest sites, such as blocking them off with screens or completely removing areas that house colonial nest sites, can reduce pigeon abundance [7]. Another strategy is removing or destroying eggs periodically. Dovecotes, structures used to house pigeon nests, have been used in cities to reduce egg production. However, maintenance of dovecotes can be time-consuming and costly, and removal of eggs may trigger pigeons to lay more [48]. Also, the use of dovecotes has not proven to be effective [18].

3.4 Chemical sterilization

Chemical sterilization is a humane strategy to decrease abundance but does not get rid of the entire population [7]. Research conducted on the effect of decreasing reproductive output through sterilization did decrease abundance in feral pigeons (among other species) and was more effective than killing individuals [49]. However, the majority of studies have been conducted in laboratory settings. How effective sterilization is in field settings with natural populations is not well understood.

3.4.1 Nicarbazin

Texas Tech University is investigating how OvoControl (Innolytics, LCC) can limit reproductive success and decrease abundance. The active ingredient in OvoControl is Environmental Protection Agency approved nicarbazin in the form of a ready-to-use bait. First developed to decrease the growth of coccidian parasites in chickens, researchers found that nicarbazin decreased egg production and

Figure 3.
Pigeon nests under a HVAC machine on top of the Biology building on TTU campus.

DOI: http://dx.doi.org/10.5772/intechopen.84612

hatchability by compromising the internal egg structure and sperm receptors [50]. When consumed by a female, the hatchability rates decreased to near zero with no other side effects [50].

In Rimini, Italy, a population management strategy was conducted to decrease pigeon abundance with Ovistop (a similar bait with nicarbazin as the active ingredient) [51]. Using 10 locations and dispensing bait from March to November in 2005 and 2006, there was a decrease in abundance by 48% [51]. The city of Modena, Italy sought to decrease pigeon abundance but wanted to make sure that native swifts and bats could still flourish. They dispensed Ovistop and decreased the number of nesting locations [52]. The scaffold holes around the city that pigeons used for nesting were not completely covered, but reduced in size to allow bats and swifts to roost inside [52]. This program led to a decrease in pigeon abundance from 1060 pigeons to 205 pigeons in downtown Modena [52].

Innolytics, LCC helped the city of San Diego study how OvoControl influenced natural pigeon populations. Two sites that housed approximately 150 individual pigeons each were selected for the study. At the first site, they dispensed OvoControl for the pigeons to consume. At the second site, they did not dispense OvoControl to see how the population progressed without the influence of sterilization techniques. Within the first year of dispensing the bait, there was a decrease in pigeon abundance by 53% [53]. After 28 months, the population decreased by 88% [53].

Texas Tech University has been working closely with Innolytics, LCC to make sure that OvoControl is used as effectively as possible. We have found it to work best on top of flat buildings, where there are hot spots in abundance. To attract pigeons to the location, we dispensed cracked corn with deer feeders. Once an adequate number of pigeons were coming to the area to consume the corn, we slowly integrated OvoControl into the deer feeders. We started this process on top of the Biology building and HH in May of 2018. We are examining how effective the treatment is by candling eggs to look at fertilization rates [54] and abundance throughout the year.

Avian birth control decreases abundance by decreasing the number of pigeons entering the population through birth [54]. Due to pigeons being social and territorial, the feeders should only attract local flocks. Once the population decreases, no new pigeons should enter the area due to individuals from the current flock still being present and defending their roosting/nesting location [55]. However, once pigeons stop eating OvoControl, it can lose its effect within 1 week [55]. By placing feeders on top of buildings, exposure to native birds is limited. There are also no secondary trophic effects. Hawks will not be affected after preying on a pigeon that has consumed OvoControl [53]. OvoControl is only useful if it is directly consumed [55].

3.5 Limiting roosting locations

Another way to decrease habitat favorability for pigeons is to decrease roosting availability. Buildings are full of ledges for pigeons to roost on, and there are many ways to decrease roost selection. Increasing the angle of the ledge to 45° or more by using sheet metal or other materials can discourage pigeons from roosting or nesting on ledges [56]. Netting or mesh can be used as well. Texas Tech University has decreased roosting and nesting by screening off balconies that housed a large colony of pigeons. Netting can be placed below rafters to discourage pigeons from roosting in barns and sheds [56].

Porcupine wires are spikes that can also be placed on ledges to prevent pigeons from roosting [56]. The stainless steel spikes point in all directions and cause discomfort for pigeons who try to roost in the location [57]. A bird shocking device

(Bird-shock, Flex-Track) can also be placed almost anywhere and has limited visibility. The device shocks the bird enough to scare it away, without harming the bird [57].

Noise-making devices such as those that emit a high-frequency sound have proven to be ineffective in removing pigeons [56]. Scaring pigeons with lights, fake snakes, owls, and other devices also seem ineffective on pigeon abundance [56]. Chemical repellents have also been used to decrease pigeon presence [56]. While effective at first, pigeons can get used to the repellent and decrease repellent effectiveness [56].

3.6 Birds of prey

Falconry is becoming a popular and "new" method in pigeon management. Using falcons may be a useful technique if there is no way to pigeon proof an area or a building [58]. Birds of prey can scare away pigeons and greatly influence their behavior [58]. In Calgary, falcons have been released at the South Health Campus [59]. Effectiveness of commercial use falconry or having falcons permanently live on buildings is not known. Increase in predator abundance within a location would be more beneficial than in areas where pigeons have very few predator encounters. Such an approach may be more beneficial than using fake owls. Pigeons can become accustomed to fake owls and no longer fear them [56].

4. Conclusion

Pigeons can influence humans through disease transmission, infrastructure damage, or financial loss. How to rid an area of pigeons is not always straightforward. Some methods are not effective in decreasing populations (**Figure 4**). Due to their rapid reproductive output, high fledgling success and homing ability, culling or trapping pigeons does not work when trying to decrease abundance. Culling and removal deems to be a short-term solution to a long-term problem. Scaring pigeons with fake birds of prey or with sound or lights also proves to be ineffective.

One of the most important aspects and the first step to any population management plan is to study the population that is a nuisance. Studying abundance, population hot spots, colonial nest locations, and primary food resources can create the building blocks of an effective plan to modify habitat and decrease carrying capacity of the location. Nest and roosting locations can be removed by modifying ledges or crevices. Decreasing the attractiveness of buildings in Perugia, Italy resulted in a 23% decrease in pigeon abundance within a year [60]. If an organization can determine the primary food source for the population, decreasing it or educating the public can prove to be very effective. Venice, Italy used to have more than 10,000 pigeons living in St. Mark's Square [18]. When they decreased the number of tourists feeding the population, the number dropped to 1,000 individuals [18]. However, if the population is flying to nonurban areas, such as agricultural fields, this method of decreasing carrying capacity by decreasing feeding sources may not work. Population models performed by Guinchi et al. found that influencing the habitat is the most effective and long-lasting method in decreasing abundance [18]. Simply killing off pigeons was least likely to control the population [18].

Guinchi et al. also discovered that controlling fertility with habitat modification may create the most profitable outcome [18]. However, when designing this plan, a city or organization needs to take into account effective budgeting. A population decrease will not happen immediately and it may take years to reach an optimal abundance. A multifaceted approach, using sterilization and habitat modification

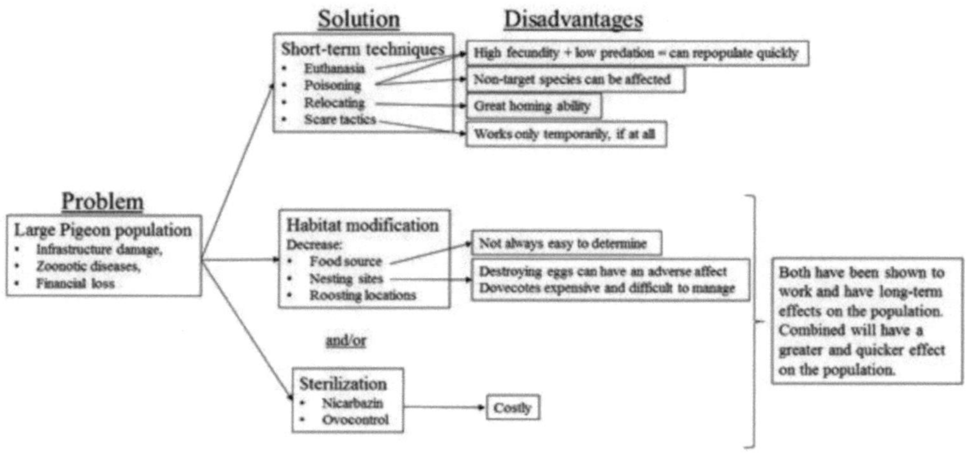

Figure 4.
Flowchart shows the most common methods and disadvantages at attempting to decrease pigeon abundance.

techniques, is useful in decreasing abundance, but can be expensive over time [7, 18]. If a sterilization technique is preferred, studying seasonal egg-laying patterns may be useful to know when to dispense nicarbazin-laced bait and when it can be suspended. Therefore, the organization is not spending money when it is not necessary.

The most effective management strategy for TTU includes habitat modification to decrease roost and nest sites, and sterilization techniques via dispensing OvoControl in locations of high pigeon abundance. Texas Tech University has decreased roosting locations using bird wire and decreased nesting locations by screening off balconies that housed a large colony. OvoControl is currently being dispensed on top of buildings that house large pigeon populations. Preliminary results from enumerating pigeon abundance on ESB and Holden Hall have shown a significant reduction within the last year. The Experimental Sciences Building is next to the Biology building where OvoControl is being dispensed, while OvoControl is being dispensed on top of Holden Hall. Bird wire has been placed on both buildings, and balconies have been screened off at ESB to decrease nesting locations. Texas Tech University is furthering their study of pigeon ecology and conducting more detailed analyses of nesting and roosting locations to further decrease habitat favorability.

Acknowledgements

We would like to thank TTU for their financial help in conducting our research. We would like to thank all volunteers for their time and efforts in researching pigeon ecology on Texas Tech campus. We would also like to express our gratitude toward Eric Wolf and Alex McDonald from Innolytics, LCC for their valuable insight on using nicarbazin bait to decrease reproductive output of pigeons on TTU.

Author details

Erin E. Stukenholtz[1*], Tirhas A. Hailu[2], Sean Childers[3], Charles Leatherwood[3], Lonnie Evans[3], Don Roulain[3], Dale Townsley[3], Marty Treider[3], R. Neal Platt II[4], David A. Ray[2], John C. Zak[2] and Richard D. Stevens[1,5]

1 Department of Natural Resources Management, Texas Tech University, Lubbock, TX, USA

2 Department of Biology, Texas Tech University, Lubbock, TX, USA

3 Operations Division, Texas Tech University, Lubbock, TX, USA

4 Texas Biomedical Research Institute, San Antonio, TX, USA

5 Museum of Texas Tech University, Lubbock, TX, USA

*Address all correspondence to: erin.stukenholtz@ttu.edu

References

[1] UN. World Urbanization Prospects: The 2015 Revision Highlights (ESA/P/WP.241). New York: United Nations, Department of Economic and Social Affairs, Population Division; 2015

[2] Hedblom M, Murgui E. Urban bird research in a global perspective. In: Murgui E, Hedblom M, editors. Ecology and Conservation of Birds in an Urban Environment. Cham, Switzerland: Springer; 2017. p. 39. DOI: 10.1007/978-3-319-43314-1

[3] Sol D, Gonzalez-Lagos C, Lapiedra O, Diaz M. Why are exotic birds so successful in urbanized environments. In: Murgui E, Hedblom M, editors. Ecology and Conservation of Birds in an Urban-Environment. Cham, Switzerland: Springer; 2017. pp. 75-90. DOI: 10.1007/978-3-319-43314-1

[4] Pimentel D, Lach L, Zuniga R, Morrison D. Environmental and economic costs of nonindigenous species in the United States. Bioscience. 2000;**50**:53-65. DOI: 10.1641/00006-3568

[5] Sacchi R, Gentilli A, Razzetti E, Barbieri F. Effects of building features on density and flock distribution of feral pigeons *Columba livia var. domestica* in an urban environment. Canada Journal of Zoology. 2002;**80**:48-54. DOI: 10.1139/z01-202

[6] Murton RK, Thearle RJP, Thompson J. Ecological studies of the feral pigeon *Columba livia* var. I. Population, breeding biology and methods of control. Journal of Applied Ecology. 1972;**9**:835-874. DOI: 10.2307/2401909

[7] Johnston RF, Janiga M. Feral Pigeons. New York City, NY, USA: Oxford University Press Inc.; 1995

[8] Sossinka R. Domestication in birds. Avian Biology. 1982;**6**:373-403. DOI: 10.1016/C2013-0-10639-3

[9] Goodwin D. Notes on feral pigeons. Avicular Magazine. 1954;**60**:190-213

[10] Gibbs D, Barnes E, Cox J. Pigeons and Doves: A Guide to the Pigeons and Doves of the World. New Haven, Connecticut: Yale University Press; 2001

[11] Jokimäki J, Suhonen J. Distribution and habitat selection of wintering birds in urban environments. Landscape and Urban Planning. 1998;**34**:253-263. DOI: 10.1016/S0169-2046(97)00089-3

[12] Ciminari ME, Del Valle Moyano G, Chediack JG, Caviedes-Vidal E. Feral pigeons in urban environments: Dietary flexibility and enzymatic digestion? Revista chilena de Historia Natural. 2005;**78**:267-279. DOI: 10.4067/S0716-078X2005000200011

[13] Sol D, Santos DM, Gracia J, Cuadrado M. Competition for food in urban pigeons: The cost of being a juvenile. The Condor. 1998;**100**: 298-304. DOI: 10.2307/1370270

[14] Rose E, Nagel P, Haag-Wackernagel D. Spatio-temporal use of the urban habitat by feral pigeons (*Columba livia*). Behavioral Ecology and Sociobiology. 2006;**60**:242-254. DOI: 10.1007/s00265-006-0162-8

[15] Goodwin D. Comparative evolocy of pigeons in inner London. Britian Birds. 1960;**53**:201-202

[16] Kösters J, Kaleta E, Monreal G, Siegmann O. Das Problem der Stadttauben. Deutsches Tierärzteblatt. 1991;**4**:272-276

[17] Spennemann DHR, Watson MJ. Dietary habits of urban pigeons (*Columbia livia*) and implications of excreta pH—A review. European Journal of Ecology. 2017;**3**:27-41. DOI: 10.1515/eje-2017-0004

[18] Guinchi D, Albores-Barajas YV, Baldacciini NE, Vanni L, Soldatini C. Feral pigeons: Problems, dynamics and control methods. In: Soloneski S, editor. Integrated Pest Management and Pest Control—Current and Future Tactics. InTech; 2012. pp. 215-240

[19] Schnurrenberger P, Hubbert W. An Outline of the Zoonoses. Ames, Iowa: Iowa State University Press; 1981

[20] Martelli P, Borghetti P. La malattia di Lyme: Una nuova zoonosis. Obiettivi e documenti veternari. 1988;**9**:15-20

[21] Haag-Wackernagel D, Moch H. Health hazards posed by feral pigeons. Journal of Infection. 2004;**48**:307-313. DOI: 10.1016/j.jinf.2003.11.001

[22] Murton R. Man and Birds. London: Collins; 1971

[23] Weber WJ. Health Hazards from Pigeons, Starlings and English Sparrows. Fresco, California: Thomson Publications; 1979

[24] Richter R, Kösters J, Kramer K. Zür paramyoxivusinfektion bei Tauben. Der Praktische Tierarzt. 1983;**64**:915-917

[25] Černík K, Tůmová B, Novotoná L, Kaminskyj B, Rajtár V. Charaterization of paramyxovirus isolated from 4001 pigeons in Czechoslovakia. Veterinary Medicine (Prague). 1985;**30**:103-117

[26] Knoll M, Kösters J, Lütticken D. Immunitätsdauer nach impfung gegen die paramyxovirose der Tauben mit einer homologen Olemulsionsvakzine. Der Praktische Tierarzt. 1986;**67**:975-979

[27] Alexander DJ, Parsons G. Pathogenicity for chickens of avian pparamyxovirus type 1 isolates obtained from pigeons in Great Britain during 1983-85. Avian Pathology. 1985;**15**:487-493. DOI: 10.1080/03079458608436309

[28] Texas Parks Wildlife Department. Public Notice on Outbreaks of Avian Paramyxovirus in Texas. 2018. Available from: https://tpwd.texas.gov/newsmedia/releases/?req=20181012a [Accessed: 2018-11-14]

[29] Martin-Albarracin V, Amico GC, Simberloff D, Nuñez MA. Impacts of non-native birds on native ecosystems: A global analysis. PLoS One. 2015;**10**:e0143070. DOI: 10.1371/journal.pone.0143070

[30] Sol D, Bartomeus I, Griffin AS. The paradox of invasion in birds: Competitive superiority or ecological opportunism? Oecologia. 2012;**169**:553-564. DOI: 10.1007/s00442-011-2203-x

[31] Toreador. Partial Annihilation Faces Tech Pigeons. 1957

[32] Butler K. Building a renaissance in west Texas. 2010. Available from: http://today.ttu.edu/posts/2010/03/campus-architecture-history[Accessed: 2018-11-14]

[33] Texas Tech Today Monthly. Texas Tech Trivia Book. 2017. Available from: http://www.depts.ttu.edu/communications/newsletter/stories/07June/tech-trivia-book.php [Accessed: 2018-11-14]

[34] Stukenholtz EE, Hailu TA, Childers S, Leatherwood C, Evans L, Roulain D, et al. A Pigeon's Eye View of a University Campus; submitted

[35] Hailu TA, Stukenholtz EE, Whibesilassie WD, Childers S, Leatherwood C, Lonnie E, et al. Seasonal pigeon abundance on urban campus linked to environmental variables. In preparation

[36] Edrich W, Keeton W. A comparison of homing behavior in feral and homing pigeons. Zeitschrift für Physik. 1977;**44**:389-401. DOI: 10.1111/j.1439-0310.1977.tb01003.x

[37] Schmidt-Koenig K. Der Einfluss experimentell veranderter Zeitschatzung auf das Heimfindervermögen bei Brieftauben. Die Naturwissenschaften. 1958;**45**:47. DOI: 10.1007/BF00635030

[38] Wiltschko W, Wiltschko R, Keeton WT. Effects of "permanent" clockshift on the orientation of young homing pigeons. Behavior of Ecology and Sociobiology. 1976;**1**:229-242. DOI: 10.1007/BF00300066

[39] Wallraff HG. Örlich und zeitlich bedingte Variabilität des Heimkehrverhaltens von Brieftauben. Zeitschrift fur Vergleichende Physiologie. 1966;**16**:513-544

[40] Gagliardo A, Ioalé P, Savini M, Lipp HP, Dell'omo G. Finding home: The final step of the pigeon's homing process studied with a GPS data logger. The Journal of Experimental Biology. 2007;**210**:1132-1138. DOI: 10.1242/jeb.003244

[41] Haag-Wackernagel D. Regulation of the street pigeon in Basel. Wildlife Society Bulletin. 1995;**23**:256-260

[42] Haag D. Lebenserwartung und altersstruktur der Strassentaube *Columba livia* forma domestica. Ornithology Beobachter. 1990;**87**:147-151

[43] Hoy MD, Bivings AE. An evaluation of controlled hunting for management of feral pigeons. In: Third Eastern Wildlife Damage Control Conference. Alabama: Gulf Shores; 1987

[44] Sol D, Senar JC. Comparison between two censuses of feral pigeon Columba livia var. from Barcelona: An evaluation of seven years of control by killing. Butlleti del Grup Catala d'Anellament. 1992;**9**:29-32

[45] Griminger P. Digestive system and nutrition. In: Abs M, editor. Physiology and Behavior of the Pigeon. London: Academic Press; 1983. pp. 19-40

[46] Murton R, Westwood N. Avian Breeding Cycles. Oxford: Oxford University Press; 1977

[47] Levi WM. The Pigeon. Sumter, South Carolina: Levi Publication Company; 1974

[48] Jaquin L, Cazelles B, Prevot-Jilliard AC, Leboucher G, Gasparini J. Reproduction management affects breeding ecology and reproduction costs in feral urban pigeons (*Columba livia*). Canadian Journal of Zoology. 2010;**88**:781-787. DOI: 10.1139/Z10-044

[49] Bomford M. A Role for Fertility Control in Wildlife Management. Canberra, Australian Capital Territory: Australian Government Publishing Service; 1990

[50] Reinoso VP, Katani R, Barbato GF. Nicarbazin reduces egg production and fertility in white pekin ducks via reducing SP3 in the perivitelline membrane. Poultry Science. 2007;**86**:536

[51] Freedom Co., Ranchio di Sarsina, Italy. Control Campaign for the Population of Urban Pigeons by the City of Rimini, Italy–2005 to 2007

[52] Ferri M, Ferraresi M, Gelati A, Zannetti G, Domenichini A, Ravizza L, et al. Control of the urban pigeon Columba livia population and the preservation of common swift and bats Chiroptera during the restoration of Ghirlandina tower in the city Modena (Italy). In: 8th European Vertebrate Pest Management Conference. Vol. 432. 2011. pp. 113-135. DOI: 10.5073/jka.2011.432.075

[53] MacDonald A, Wolf E. Ovocontrol P 0.5% (nicarbazin) population dynamics in pigeons. In: [Poster] 6th International IPM Symposium. Oregon: Portland; 2009

[54] Ernest R, Bradley F, Delany M, Abbott U, Craig R. Egg candling and breakout analysis. ANR Publication. 2004;**8134**:1-9. DOI: 10.3733/ucanr.8134

[55] Innolytics, LLC. Ovocontrol. 2018. Available from: https://ovocontrol.com/ovocontrol/ [Accessed: 2018-11-14]

[56] Williams DE, Corrigan RM. Pigeons (rock doves). In: The Handbook: Prevention and Control of Wildlife Damage. Vol. 69. 1994. pp. 1-11

[57] Bird Barrier America. Bird Barrier. 2018. Available from: https://birdbarrier.com/ [Accessed: 2018-11-15]

[58] NBC environment. Falconry response bird control. 2018. Available from: https://www.nbcenvironment.co.uk/bird-control-services/bird-deterrents/falconry-response/ [Accessed: 2018-11-15]

[59] CBC. Calgary hospital deploys falcons to chase off pesky pigeons. 2017. Available from: https://www.cbc.ca/news/canada/calgary/hospital-falcons-hunting-pigeons-1.4198194 [Accessed: 2018-11-15]

[60] Ragni B, Velatta F, Montefameglio M. Restrizione dell'habitat per il controllo della poplazione urbanadi Columba livia. In: Anonymous, editor. Cotrollo delle popalazioni ornitiche sinantropiche: problem e prospettive. Roma; 1996. pp. 106-110

Chapter 4

Cheetahs Race for Survival: Ecology and Conservation

Laurie Marker

Abstract

Cheetahs reach speeds of up to 113 km/h accelerating from zero to 96 km/h in 3s. Revered for 5000 years throughout Asia, Europe and Africa has contributed to the species decline. Today's wild cheetah population is estimated at 7100 adult and adolescents, a 90% reduction from a century ago, and a range reduction of 9%. Over 80% live outside protected areas where human-wildlife conflict occurs. Female cheetahs live solitarily with their cubs; male cubs form lifelong coalitions. Living in low densities cheetahs' home ranges cover over 1500 km^2, requiring large landscapes with prey. Although cheetahs' lack genetic diversity from a historic population bottleneck, their greatest conservation problems are humans. Habitat loss and declining preybase leads to conflict with livestock farmers. Additionally, illegal wildlife trafficking of cubs is affecting small populations in the Horn of Africa. Solving the cheetah conservation crisis is critical and involves addressing a complex web of social, environmental and economic issues, and depends on a holistic approach balancing the needs of humans and cheetahs sharing land. Research into conserving and restoring habitat for cheetahs includes training, the use of Livestock Guarding Dogs, and other conflict mitigation strategies, addressing habitat loss, dismantling the illegal pet trade, and encouraging coexistence.

Keywords: cheetah, predators, protected areas, livelihood development, conservancies, illegal wildlife trade

1. Introduction

The cheetah, *Acinonyx jubatus*, is one of the oldest big cat species, with ancestors that can be traced back more than five million years to the Middle Pliocene era [1–3]. The cheetah is also the world's fastest land mammal, an icon of nature, and the most unique of the 41 species of cats [4]. The hunting style of the cheetah—to swiftly pursue prey over a moderate distance—necessitates some morphological and physiological specializations for both prey-killing and locomotion [5].

As an animal built for speed, all parts of its body have evolved for precision and agility. Because of its small, aerodynamic head, lean body, long legs, flexible backbone and tail that works like a boat's rudder, the cheetah can change direction in a split second and reach speeds of up to 113 km/h while turning 180° [6–8]. With each stride, the cheetah covers 6 m with just one foot touching the ground at a time; at two points in the stride, all four feet are in the air. The cheetah's flexible spine acts like a spring as it doubles up with feet under its body to clench the earth with powerful, semi-non-retractable claws, thrusting it forward with great speed and maximum distance. The cheetah is not only the fastest running land mammal; it is also known for its rapid acceleration, as it can go from zero to 96 km/h in just 3 s [6].

With less than 7100 adults and adolescents remaining [9], the cheetah is one of the most endangered big cat species. Cheetah numbers have declined primarily due to increased human-wildlife conflict, loss of habitat and loss of prey, and the illegal wildlife trade. In addition to these threats, cheetahs lack genetic variation due to a historic population bottleneck, approximately 12,000 years ago, which makes the cheetah more vulnerable to ecological and environmental changes [10–12].

Today, nearly 80% of the remaining world's cheetahs are found outside of protected areas living near rural livestock farming communities [9]. Protected areas, such as wildlife reserves or national parks typically have higher densities of larger or more aggressive predator species that can outcompete cheetahs, making it difficult for cheetahs to survive. Despite being one of the best hunter species on the savanna, cheetahs often lose their kills to larger predators. In protected areas, cheetahs have been found to lose 10–15% of their kills to lions (*Panthera leo*), leopards (*Panthera pardus*), jackals (*Canis aureus*), and hyenas (*Hyaenidae*) [13, 14]. In addition to hunting pressures, cheetahs face direct threats by larger carnivores that may try to kill an adult cheetah or its young, to reduce competition for prey and territory [13–17].

Living outside protected areas prevents threats by other predators but puts the cheetah in direct conflict with commercial and subsistence livestock farmers [18, 19]. These farmers often perceive cheetahs to be a threat to their livestock, which leads into economic and emotional issues. The Rangewide Cheetah and Wild Dog program, an IUCN Cat Specialist endorsed program, brings together conservation organizations across the cheetah's range to work on a more sustainable future for cheetahs and farmers. Cheetah Conservation Fund (CCF), Cheetah Conservation Botswana (CCB), and the Ruaha Carnivore Project work with other stakeholders, such as community members, local and national governments, conservancies and scientists to develop and implement action plans for cheetah conservation throughout its range [9, 20–23].

As human populations grow, so do the chances of conflict with cheetahs. Simultaneously, available rangeland will shrink, along with the wild prey base, hastening the decline of the cheetah [24, 25]. If the observed trends of decline among cheetah population continues, the world's fastest land mammal could become extinct within the next 15–20 years [26].

2. Cheetah distribution

The cheetah was once one of the most widely distributed of all land animals. Through the course of time the cheetah was found from North America to China, throughout Asia, India, Europe, and Africa. About 20,000 years ago, it settled into its current range [3, 27].

A century ago, approximately 100,000 cheetahs were found in at least 44 countries throughout Africa and Asia. Today, the current free-ranging populations of cheetahs are restricted to 10% of their former range, found only in small, fragmented areas spread across 23 countries in Africa (in North Africa, the Sahel, East Africa and southern Africa), however, two thirds of these countries' cheetah populations number less than 200 individuals [9, 28]. It is estimated that fewer 50 wild cheetahs remaining in Iran, the last of the Asiatic population [9, 29].

Today, viable populations may be found in less than half the countries where cheetahs still exist. Cheetahs are particularly difficult to census due to their large home ranges, which average more than 1500 km^2 [14, 30–32], and their shy nature, an instinct that has been reinforced because of persecution on farmlands, where

DOI: http://dx.doi.org/10.5772/intechopen.82255

they are shot, trapped and chased [19, 22, 33, 34]. As a result of persecution and due to their naturally large home ranges, wherever they live they occur in low densities [34].

All populations of cheetahs are listed on the Convention on International Trade in Endangered Species of Fauna and Flora (CITES) Appendix 1 and are classified as Vulnerable or Endangered by the International Union for Conservation of Nature (IUCN) [35]. All cheetah populations are threatened due to habitat reduction and declines in prey populations, which bring them into increased contact (and ultimately conflict) with farmers and livestock [14, 20–23, 33].

Due to its declining numbers and genetic lack of diversity, it is important to protect remaining wild cheetah populations to ensure the species chances for survival. An evaluation of conservation priorities in each country where the cheetah is found has been conducted to better understand the issues involved in achieving this goal [20–23, 33, 36]. The remaining strongholds for cheetahs are Namibia and Botswana, in southern Africa; and Kenya and Tanzania in East Africa (see **Figure 1**). With approximately 20% of the world's remaining wild cheetahs and successful efforts to conserve its wild population, Namibia is popularly known as "The Cheetah Capital of the World."

As a result of habitat fragmentation over time, there are currently four genetically confirmed subspecies of cheetah, three African and one Asiatic subspecies [4, 12, 37]. These subspecies are physically distinct from one another, and research is still ongoing to determine the genetic uniqueness of each. One previously-accepted subspecies, the Northeastern African Cheetah, *Acinonyx jubatus raineyii,* which was found in Kenya, Tanzania, and Uganda, was determined in 2017 to be a conspecific of *A.j. jubatus* in 2017 and reclassified as such [4]. The currently classified subspecies of cheetah are as follows:

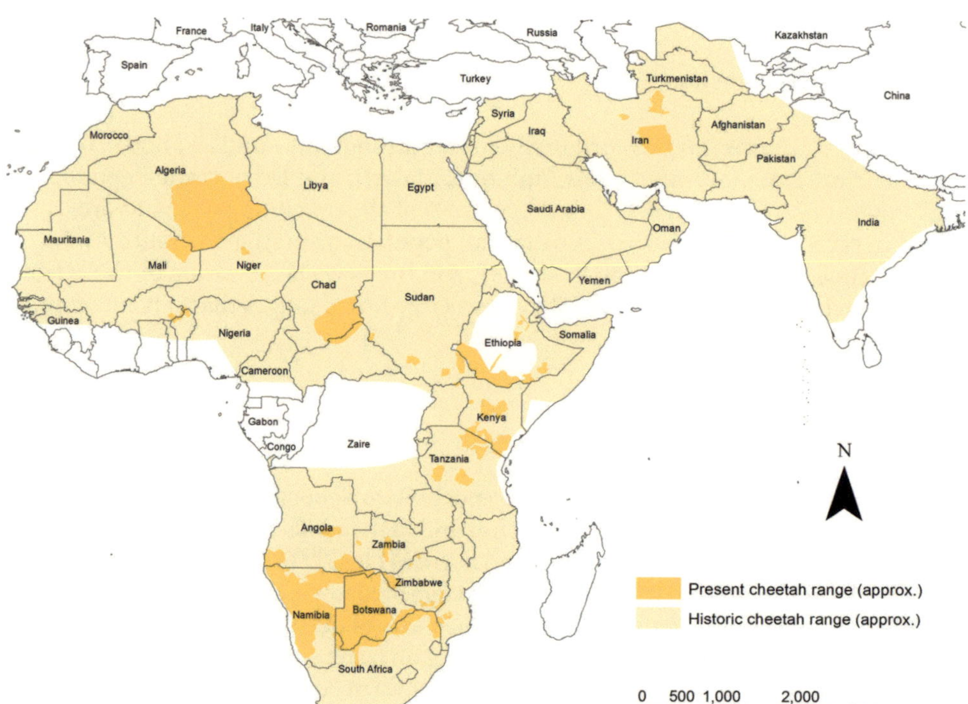

Figure 1.
Historic and current: cheetah range map [28].

2.1 Asiatic cheetah

Acinonyx jubatus venaticus originally found throughout Asia in Afghanistan, India, Iran, Iraq, Israel, Jordan, Oman, Pakistan, Saudi Arabia, Syria and Russia [38], but now there are approximately 50 remaining in a small, fragmented populations in Iran [29]. These cheetahs have denser fur and what appears to be a mane, which is actually extra tufts of hair on their neck and shoulder blades. This thicker coat keeps them warm in the cooler nighttime temperatures of their environment.

2.2 Northwestern African or Saharan cheetah

Acinonyx jubatus hecki, historically found in Northwest Africa in Egypt, Algeria, Tunisia, Morocco, Niger, Mauritania, Mali and in some western African countries, including Benin, Burkina Faso, Senegal and Ghana [38]. Today, small populations remain in Algeria, Niger, Benin, and Burkina Faso [28]. Most are in the central western region of the Saharan desert and the Sahel. Unique in appearance, this subspecies in critically endangered with only approximately 250 individuals remaining. Saharan cheetahs are extremely pale, almost white in color. Their coloring provides them with a natural defense against detection in the desert environment. Sometimes there are no spots on its face and its tear marks appear to be missing. In general, they are smaller than the other subspecies. This may be a product of their desert environment, where there is less prey for them to hunt.

2.3 Eastern African or Sudan cheetah

Acinonyx jubatus soemmeringii, was historically found in Sudan, Djibouti, Cameroon, Niger, Nigeria, Chad, Ethiopia and Somalia [38]. Today, small populations are in Chad Ethiopia, Somali, Somaliland, and South Sudan [28]. This subspecies is the richly colored.

2.4 Southern African cheetah

Acinonyx jubatus jubatus, originally found throughout Namibia, Botswana, South Africa, Zambia, Angola, Mozambique, Malawi, and Democratic Republic of Congo and Zimbabwe [38]. Today, over 50% of the remaining cheetahs are found in Namibia and Botswana, with small populations in Angola, South Africa Mozambique, Zambia and Zimbabwe [28], and the new classification of those from Kenya, Tanzania and Uganda [4]. They have large home ranges that can encompass hundreds of square kilometers.

3. Cheetah behavior

Cheetahs have evolved for speed and are not built to fight other predators. Because of this, they are shy in nature and will often abandon their kills in the presence of more robust predators, such as lions, leopards and hyenas [14, 39]. To combat this, cheetahs are typically diurnal hunters, as opposed to other large predator species, such as lions, hyena, and leopards, which are nocturnal [39, 40]. Their lack of defense against these predators has led to 80% of the current cheetah range being on farmland habitat [28, 41].

Female cheetahs live solitary lives and do not form coalitions. After a 93 to 95-day gestation, solitary female cheetahs give birth to two to six cubs, with 3.5 being the average litter [13, 28, 42]. Cubs stay in the den for the first 6 weeks, with females moving their cubs to different nest sites for protection [39, 43].

DOI: http://dx.doi.org/10.5772/intechopen.82255

At 6 weeks of age, the cubs leave the den and follow their mother. At first, cubs will stay hidden while their mother begins to stalk prey. While the cubs are on their own, they may chase after inappropriate prey animals, such as jackals or birds [13]. To teach them how to catch appropriate prey, their mother will capture and release prey for the cubs to play with to practice their hunting skills [13]. The cubs will begin to chase the prey and trip it before the mother eventually comes and kills it for them [43]. Cubs start initiating their own hunts at about 1 year of age but are not proficient until they are independent [13]. Cubs stay with their mother for about 18–22 months. Even after they become independent, it can take cubs up to 3.5 years to master hunting [13]. Female cubs establish their home ranges within their mother's larger home range, so there is familiarity among female cheetahs that are related [44].

Once the young males disperse they will not maintain a territory until they are 4–5 years of age [13, 18]. Male cheetahs remain with their other male siblings from birth, forming a coalition for life. This behavior increases hunting success and is a defense against predators. By sticking together, they can hold and defend a better territory, where wildlife prey is abundant [39]. This also increases the chances of a greater number of desirable females coming into the territory for breeding.

Members of a coalition are very bonded to one another. If separated, they do a lot of vocalizations. Cheetahs have a variety of unusual vocalizations including a dog-like bark and a bird-like chirp for calling between each other [13, 39]. Other sounds they make include a bubble or "uhun" sound, a hiss, and a growl. They are also very affectionate to each other. They purr and lick each other's faces. Male coalitions work together when hunting and are able to hunt larger prey together. Cheetah coalitions are very stable, and the bond of brotherhood is permanent.

4. Cheetah ecology

Cheetahs require vast expanses of land with prey and other resources [9, 45]. Research in Namibia shows that cheetahs have an average home range of 1500 km^2 with individuals covering 20–40 km in a week, but live in low densities throughout their range [14, 39, 45, 46]. Most cheetahs live in open grasslands and savanna, which are arid environments [35]. Throughout the cheetahs range, cheetahs are known to use tall trees for greater visualization as well as territorial markings. In Namibia, these trees have been called "playtrees," as cubs are often seen climbing into them, or "newspaper tree," as male cheetahs use these trees for leaving their territorial scent marks, urine and feces [14, 47]. However, in many ecosystems throughout the cheetah ranges, bush encroachment, a form of desertification caused by overgrazing arid landscapes as well as the decline of many of the large mega-herbivores, has caused a problem for cheetah hunting ability as well as altering the mix of wildlife [47]. Bush encroachment results in the prolific growth of a native plant species, *Senegalia* sp., commonly known as thornbush [47]. On traditionally open savanna lands where cheetahs hunt using their natural advantage, speed, bush encroachment changes the landscape. Increased bush limits the cheetahs' vision and speed, which lowers their hunting success, ultimately altering the mix of wildlife [47, 48]. Decreased vision does not just hinder their hunts, as cheetahs are also more likely to consider livestock as prey over wildlife, becoming a problem animal to farmers and increasing conflict [49]. Throughout Africa, cheetahs are known to frequent certain trees [14, 47].

Changes in the arid ecosystems in favor of human needs have also created problems, mainly from overgrazing of livestock leading to desertification, leaving limited grazing for wildlife. Further compounding this issue, forces of nature that are unpredictable and difficult to manage such as climate change, negatively affect agriculture and wildlife as rangelands become drier and vegetation is altered [50]. This also affects distribution and abundance of prey [14, 45]. And, as the human population grows, air and water become more polluted, habitat is lost to development, and the climate crisis deepens. Ultimately, the cheetah's chances for survival depend greatly on the re-balancing of the ecosystem and the restoration of habitat so it will support sufficient natural prey [26].

Learning to hunt is the most critical survival skill that the cubs must develop [13]. At 1-year-old cubs are participating in hunts and the mother, while assuring enough kills for the family's survival, will allow the cubs to join in. Cheetahs hunt in the early morning and early evening and capture their prey by stalking to within 10–30 yards or as far as 80 yards before beginning the chase [13]. During a hunt, cheetahs usually catch their prey after an average 200-yard sprint [13]. Although fast, their ability to accelerate at a high speed is most critical, and their maneuverability enabling them to turn rapidly is more important than their speed. Most hunts take place at a slower speed, as prey are dodging in efforts to flee [39]. Successful hunters need not only speed but stealth as well. They move slowly and remain low in the grass, staying downwind, sometimes hiding behind small mounds to obscure their approach, taking advantage of their coloring to camouflage their appearance and blend into their surroundings [13].

Only 10% of cheetah chases are successful, and diet depends largely on where the cheetah lives [13]. Medium-sized and smaller prey, such as antelope and gazelles, hare and the young of larger antelope like wildebeest (*Connochaetes*), kudu (*Tragelaphus* sp.) or oryx (*Oryx* sp.) and small warthog (*Phacochoerus*) are the most common targets, and coalition males often take larger prey species like zebra (*Equus hippotigiris, E. dolichohippus*), kudu or ostrich (*Struthionidae*) [14]. Asiatic cheetahs prey on goitered gazelle (*Gazella subgutturosa*), ibex (*Capra sp.*), wild sheep (*Ovis* sp.) and chinkara (*Gazella bennettii*) [14]. Factors that lead to a successful hunt include herd size, prey response, number of cheetahs hunting, and the distance the cheetah runs [13, 39]. The cheetah can go days without water because they get hydration they need from the blood of their prey and will gorge themselves on a big enough kill and then fast for 2–5 days; however, they will hunt daily if possible [13].

Threats from other predators is one of the main reasons why nearly 80% of wild cheetahs today are found outside of protected areas (like national parks or wildlife reserves) and living alongside human communities [9, 28, 39]. In protected areas, cheetahs often lose their kill to larger and more aggressive predators. Cheetahs tend to lose 10–15% of their kills to other predators [39]. Cheetahs are apex predators and the best hunters on the savanna, they feed many species with their kills' thus increasing biodiversity of the ecosystem in which they live [14]. Without this balance, other species within the ecosystem will also be adversely affected, ultimately resulting in negative consequences for the human population.

5. Threats to the cheetah

Conservation research shows that the greatest conservation problems are not biological but have more to do with humans. Climate change and human population growth compounds these threats to an already genetically compromised species [19, 25, 33, 35, 45, 50]. Human-wildlife conflict, habitat loss and illegal wildlife trade have become the biggest threats to long-term cheetah survival [9, 26, 51, 52].

The majority of people who live alongside cheetahs are rural subsistence farmers whose livelihoods depend on the health and wellbeing of their livestock. These farmers have traditionally viewed cheetah as worthless vermin, a nuisance and a

DOI: http://dx.doi.org/10.5772/intechopen.82255

threat. Some governments have sanctioned herd protection programs that allow for cheetahs on farmlands to be trapped and removed or killed on sight [51]. Culling of cheetahs in Namibia during the 1980s resulted in losses of nearly 7000 cheetahs due to real and perceived conflict with livestock and game farmers [18, 53]. While these programs were popular during the 1970s and 1980s, this led to a rapid, widespread reduction in the numbers of wild cheetah, which fortunately has been stemmed by the intervention of conservationists and the introduction of non-lethal predator control techniques [19].

In Northern Africa, the rarity of the Saharan or desert cheetah is directly linked to the rarity of the prey species, as the IUCN Red List lists both predator and prey as critically endangered species [28]. The Saharan cheetah can still be found in small numbers in Algeria (Ahaggar and Tassili N'Ajjer), Niger (Termit and Aïr), and possibly also in Mali, Chad and Mauritania [9, 28]. Due to the decline of prey, mainly from poaching and overhunting, these cheetahs are living primarily on hare (*Lagomorpha*) but are known to attack and kill young camels (*Camelus dromedarius*) and goats (*Capra sp*), provoking retaliation [45].

Habitat destruction across Africa and Iran is one of the biggest problems threatening cheetah survival. As wild lands are being destroyed and fragmented by human expansion, landscapes across Africa that once supported thousands of cheetahs now support only a few. With habitat loss comes the decline in wild game species that provide prey for the cheetah [45]. As the human population continues to grow exponentially, there is an every-increasing demand for land rights. This affects the cheetah, as increased agricultural pressure and subdivision of land mean a decrease in available habitat for the cheetah and other wildlife species [25].

For many African wildlife species, living within a protected national park or private game reserve such as the Maasai Mara in Kenya, the Serengeti National Park in Tanzania, or Kruger National Park in South Africa is the difference between life and death. Animals that live on protected lands are guarded by rangers and photographed by tourists, which makes them less likely to be poached. But for some species, including the cheetah, living in protected areas results in greater competition with other larger and more aggressive predators that will steal their kills and kill their cubs. There is a high cub mortality (up to 90% in protected areas), mainly due to predation [15]. Consequently, nearly 80% of all cheetahs throughout their range are found living outside of protected parks and reserves [9, 28].

The lifespan of an adult cheetah is between 8 and 10 years [18, 42]. Adult mortality is one of the most significant limiting factors for cheetah population growth and survival [18, 54, 55].

Because of in-depth, *in situ* research studies of the wild cheetah that have taken place since the early 1980s, we probably know more about this species than most any other big cat species [24]. The unique genetic profile of the cheetah demands a thorough understanding of their biology and capacity for reproduction. A potentially critical factor for the long-term survival of the cheetah is its lack of genetic variation relative to other felids [12].

Genetic homogeneity can make a species more susceptible to ecological and environmental changes to which the world is subjected now and has been interpreted in the context of two potential risks: the expression of recessive deleterious alleles and increased vulnerability to viral and parasitic epizootics that can affect genetically uniform populations [11, 12]. Cheetahs are known to be very susceptible to several feline diseases and are possibly more vulnerable due to the lack of heterogeneity in the population [11, 56, 57]. As cheetahs transverse the farmlands where more villages occur, the potential for disease transmission increases. Given the species lack of genetic diversity, monitoring the overall health of cheetah populations is an important component of understanding and promoting its long-term viability.

Another major threat is the trafficking of live cheetahs for the illegal pet trade. Wildlife trafficking is one of the top five transnational crimes and it is impacting affecting the survival of many species (U.N. Office on Drugs and Crime). While cheetahs are not poached at the same high rates as elephants and rhinoceros in Africa, an estimated 300 cheetah cubs are being smuggled out of the continent each year to supply the illegal pet trade [52] (**Figure 2**). Illegal capture is occurring mostly in Ethiopia, Somalia and northern Kenya, with most cases being reported in Somaliland [52]. Although trade in wildlife species products is regulated by both international and national laws, the illegal wildlife trade is estimated to be worth between $50–150 billion USD annually. Cheetahs, listed as an Appendix 1 species under CITES, are removed from the wild for the pet trade and for their body parts.

Because the cheetah is light and built for speed and has a flight versus fight instinct. For this reason, the cheetah is a sought-after pet in multiple regions of the world [52]. In the Gulf States, cheetahs are one of the most popular exotic pets and are a status symbol [52]. Photos posted on social media show cheetahs with gem-studded collars posing in luxury vehicles beside their owners, or riding in speed-boats, or in other outlandish depictions.

Keeping a wild cheetah as an exotic pet undermines the species, as its numbers are so low it cannot sustain regular losses and still hope to survive. The illegal pet trade is decimating cheetah populations that are already small and nearly unsustainable [9, 52]. Five out of six cubs poached die before being sold into the pet trade. Cheetah cubs that survive long enough to be sold most likely will not make it beyond 2 years of age. All will become sick, disabled and die prematurely. Improper diet, environment and lack of veterinary care result in a myriad of debilitating health problems [52].

Another human issue impacting the cheetah is tourism. Everyone who visits Africa on safari wants to see a cheetah. While tourism helps bring international attention to the cheetah and instills economic value in species survival, crowds of multiple vehicles surrounding cheetahs can have a negative impact [20, 58, 59]. Cheetahs hunt in the early morning and late afternoon when most game drives take place. Vehicles sometimes move between the cheetah and its prey so tourists can

Figure 2.
A caged cheetah cub confiscated from illegal wildlife trafficking. Cheetahs are often illegally sold as pets to the Middle East and for everyone that makes it live into the trade, 5 die in transport.

get a better view. This interferes with the cheetah's ability to catch its prey and can separate mothers from cubs [20, 58, 59].

Predators are exceptionally aware of tourists and their vehicles and sometimes use them to their advantage. If a cheetah has made a kill it will most certainly lose it if vehicles are present, since other predators, particularly the hyena, lion or jackal are alerted by the tourists. If the cheetah has cubs, this is a very dangerous situation for them, as they are made more vulnerable by the interference of the vehicles. Research conducted in the Maasai Mara recorded that nearly 30% of cheetah sightings had more than 20 vehicles surrounding it, and of these, more than 50% were less than 30 yards from the animal [58]. Nearly 60% were reported as being noisy (hooting and engine revving) with tourists and drivers shouting or talking very loudly [20]. The busiest time for the tourist vehicles was found to be between 4:40 and 6:30 pm coinciding with the high times for hunting by cheetahs [20].

In the Maasai Mara, a high incidence in sarcoptic mange in cheetahs has been linked to stress caused by tourism vehicles. Chronic stress induces immunosuppression, which in cheetahs has been found to contribute to a high occurrence of uncommon diseases, like mange, gastritis and amyloidosis [56–58, 61].

6. Conservation initiatives

Solving cheetah conservation crisis involves addressing a complex web of social, environmental and economic issues. Although people are the root of most of the problems facing the cheetah in today's world, they are also the solution as well. Over the past several years, conservation professionals have come together to look closely at the crisis for the cheetah and devise strategies for cheetah survival [9, 24].

Though the situation for the wild cheetah is dire with less than 7100 wild adult and adolescents remaining, there is hope for species' long-term survival. Efforts to educate communities living alongside cheetah through awareness building media campaigns and to obtain government buy-in have been successful [60]. Range-wide strategies for the cheetah have been developed and implementation is underway through eastern, southern, north, west and central Africa [9, 14, 36]. Capacity building for range country conservation scientist and agriculture extension officers is an ongoing process, using the "train the trainer" approach [26, 60]. Committed conservationists are focusing on the bigger picture, encouraging community participation in finding solutions that alleviate conflict. The bigger picture allows for a global perspective and a multi-species, integrated approach to cheetah conservation.

At the 2003 World Park's Conference, conservation practitioners agreed there was need for community ownership and responsibility over assessing and addressing human wildlife conflict (HWC). To be successful, improved communication on a local level between stakeholders and on a global level between experts, practitioners, local communities and international conservation organizations would be required. Guidance manuals, processes and systems needed to be developed, and HWC mitigation needed to be supported by international political and legal institutions.

In 2007 and again in 2012, government representatives, non-government organizations (NGOs) and the International Union for Conservation of Nature (IUCN) Species Survival Commission's (SSC) Cat Specialist Group met to develop regional strategies for the survival of cheetahs [20–22]. Since then, strategies for the three regions of Africa have been developed: central, west, and southern, eastern and north. The Range Wide Conservation Plan is a joint initiative of the Wildlife Conservation Society and the Zoological Society of London, in partnership with the

Cat and Canid Specialist Groups of the IUCN/SSC. These strategies have created a structure under which government programs could be developed, thus enabling conservation action on a national level. Subsequently, National Action Plans have been developed in 13 cheetah range countries [14, 20–23].

Today, cheetah research and conservation programs are found in Botswana, Iran, Kenya, Namibia, Tanzania, South Africa, and Zimbabwe [14]. Furthermore, cheetah research and training has been conducted in countries such as Algeria, Angola, Benin, Ethiopia, Mozambique, Niger, Zambia and Somaliland [26].

Community-based, natural resource management NGOs are also working with many communities throughout Africa to develop integrated programs incorporating tourism development and economic incentives to diversify livelihoods for its citizens [62]. Through outreach programs focusing on agricultural education, farmers are being taught about livestock health and management along with grasslands, wildlife and basic principles of ecology [19, 60]. Conservancies—collaborative partnerships of neighboring farms united by common operating principles—are being formed to implement standardized land management techniques that benefit people, livestock and wildlife [33, 49, 63, 64]. Examples of successful conservancies are being used to provide the basis for developing large-scale trans-boundary land management plans for the future [64].

Conservation biologists increasingly underscore that national parks and reserves alone are not large enough to sustain the wildlife they were created to protect. This is particularly true for the cheetah [25, 65]. Therefore, the focus on conservation of private land is crucial. Conservancies are one of the most important solutions for cheetah survival as they promote sustainable management of natural resources and development of responsible eco-tourism [64]. Conservancies give communities a vested interest in the welfare of local wildlife by giving them control over the economic benefits from wildlife populations. As a result, fewer problems with poaching are experienced and human-wildlife conflict is reduced [49].

With populations dwindling through most cheetah-range countries, cheetah survival depends on people using an informed, integrated approach to conservation. Education is the foundation and must include communication, information sharing and capacity building [60, 66]. In 2005, CCF began conducting month-long courses to bring together conservation managers, scientists, and community representatives from African cheetah-range countries and Iran [66]. The courses build capacity, with a goal of stabilizing cheetah populations. More than 300 participants are now managing cheetah and wildlife conservation programs in their own countries.

At the same time, research into ways to conserve and restore habitat for cheetahs and farmers is also important by working with local livestock farming communities, to help improve their livelihoods. Assigning economic value to cheetahs and having a thriving population on the landscape is key. Training programs have been developed by the Cheetah Conservation Fund that address human-wildlife conflict called Future Farmers of Africa (FFA) [66]. FFA teaches best agricultural practices to rural farmers to help them manage integrated wildlife and livestock farmlands. FFA also teaches how non-lethal predator control methods can reduce predation losses. The use of livestock guarding dogs is included in this course. CCF has helped develop similar programs throughout the cheetah's range. Many of these methods of reducing predator conflict are also applicable or adaptable to other animals such as mountain lions, jaguars and wolves, and have been used as models elsewhere in the world.

FFA covers topics like livestock health, veterinary care, husbandry, and valuation as well as wildlife and rangeland management, methods of non-lethal predator control, predator identification and best practices to reduce livestock losses including the use of kraals, birthing camps as well as seasonal, coordinated breeding. The use of a livestock guarding dog has been shown to be a very effective tool and is included in training [67] (see **Figure 3**). The Anatolian shepherd or Kangal dogs have been used for thousands of years in the Turkish region of Anatolia as livestock guarding dogs, where they were formidable guardians of livestock against bears and wolves [69–71].

Since 1994, the Cheetah Conservation Fund (CCF) in Namibia, has bred and placed these dogs with livestock farmers to reduce conflict with livestock and reduce the killing of cheetahs and other predators. Farmers who use CCF LGDs report a decrease in predation rates ranging over 80% [70, 72]. Simultaneously, LGDs reduce the killing and capture of cheetahs and other predators [72, 73]. The dogs have been so successful, similar programs in South Africa, Botswana and in Tanzania [68].

Increasingly, today's consumers rely on product labels to guide their purchases, and at the same time, are willing to pay a premium price to ensure a product's providence. In 2000, CCF conceptualized the Cheetah Country—Eco-Labeling Program to encourage predator-friendly farming techniques in producing beef, goat cheese, crafts, honey and wine [49, 62]. Under the brand *Cheetah Country*, CCF hopes to transform the perception of cheetahs from vermin that threaten farmers' livelihoods into that of a precious natural resource fostering tourism and economic development. The eco-label certifies a product meets or exceeds a set of consistent standards for environmental protection or social justice. Cheetah Country Beef, the eco- label for cattle farmers who ascribe to predator-friendly farming practices, has not yet been launched, however, under the voluntary certification, farmers would sign an agreement stating they will not indiscriminately kill cheetahs on their farmland and in return, they would receive a price premium for their meat [49, 62, 74]. The extra money will help farmers cover the cost of implementing non-lethal predator control measures, like the cost of calving kraals or keeping a livestock guarding dog [49]. The most successful example of an eco-label in food production is dolphin-friendly tuna, a concept that has gained traction around the world. Wildlife Friendly Enterprise Network (WFEN), an international organization that promotes conservation through facilitation and certification of responsibly-produced agricultural products is putting these concepts to use for cheetahs as well, if WFEN's standards for certification are met [62].

Figure 3.
A goat herd protected by a livestock guardian dog in Namibia. Turkish Anatolian shepherd and Kangal dogs are bred and placed with livestock through the Cheetah Conservation Fund in Namibia.

7. Conclusion

The outlook for the cheetah is today in human's hands. Cheetah populations can rebound. But humans also have the capacity to save them. In many parts of Africa, cheetahs and other large predators are viewed as threats to human livelihoods, rather than species vital to maintaining healthy, balanced ecosystems. Good livestock management can protect herds while allowing prey and room for cheetahs and other predators. Having thriving cheetah populations also brings economic value to land as they and other predator species help drive tourism.

Implementation of more programs now is critical, so future generations will benefit from having cheetahs on earth. Continuing to expand our scientific research will be important (**Figure 4**), while collaborating with international institutions in fields such as cheetah health, genetics, reproduction, ecology to establish population numbers, as well as expanding training and capacity building programs will be key in cheetah conservation, while expanding efforts to stop the illegal cheetah trafficking. If we wait much longer, we will lose this amazing feline icon of speed and grace. A holistic approach that considers all stakeholders is critical to balance the needs of people, wildlife and the land and try to make their efforts sustainable. This way, the communities are more likely to be good stewards of wildlife. The end goal to save the cheetah is to achieve coexistence. This is the only way to ensure a permanent place for cheetahs on Earth.

Education and outreach are key in building awareness for the cheetah's plight and for developing sustainable practices that alleviate pressure on the species. Looking to the future, teaching conservation and instilling a high regard for the environment among young learners will help cheetahs secure a permanent place on Earth.

Creative approaches are also necessary. The future of the cheetah will require enhancing the livelihoods of the human communities that live alongside them. These include developing alternative income sources, such as eco-tourism, economic incentives for predator-friendly products. The concept is that farmers in cheetah range areas can be monitored and certified as practicing predator-friendly livestock management. In return for being good stewards to the cheetahs on their land, these farmers can be certified with the Cheetah Country eco-label and receive premium prices for their products [49, 62]. A program in development, its model

Figure 4.
Satellite collars allow monitoring of cheetahs movements. Through understand the cheetah's use of their large home ranges (ave. 1500 km^2) allows for management plans can be used with rural communities to plan for the cheetahs' survival in the future.

could serve to protect all of the world's predators, each of whom are threatened by conflict with humans and yet are vital to maintaining the health and biodiversity of their ecosystem.

Despite all of the problems facing the cheetah, including genetic uniformity, competition with other large predators, destruction of habitat and conflict with humans, this iconic animal has survived for thousands of years. Cheetahs continue to fulfill their ecological role as the fastest mammalian apex predator on land. With integrated conservation programs across large landscapes, survival of cheetahs for future generations can be attained [26].

Acknowledgements

Thanks to the Cheetah Conservation Fund for their support of long-term research (www.cheetah.org), Susan Yannetti and Natalie Minor for their assistance with editing of this Chapter.

Author details

Laurie Marker
Cheetah Conservation Fund, Otjiwarongo, Namibia

*Address all correspondence to: director@cheetah.org

References

[1] Johnson WE, Ezirik E, Pecon-Slattery J, Murphy WJ, Antunes A, Teeling E, et al. The late Miocene radiation of modern Felidae: A genetic assessment. Science. 2006;**311**:73-77. DOI: 10.1126/science.1122277

[2] Li G, Davis BW, Eizirik E, Murphy WJ. Phylogenomic evidence for ancient hybridization in the genomes of living cats (Felidae). Genome Research. 2016;**26**:1-11. DOI: 10.1101/gr.186668.114

[3] Van Valkenburgh B, Pang B, Cherin M, Rook L. The cheetah: Evolutionary history and paleoecology. In: Marker L, Boast LK, Schmidt-Küntzel A, editors. Biodiversity of the World—Cheetahs: Biology and Conservation. 1st ed. San Diego: Elsevier; 2018. pp. 25-32

[4] Kitchener AC, Breitenmoser-Würsten C, Eizirik E, Gentry A, Werdelin L, Wilting A, et al. A revised taxonomy of the Felidae: The final report of the cat classification task force of the IUCN cat specialist group. Cat News Special Issue. 2017 January;**11**:18-28 Available from: https://repository.si.edu/bitstream/handle/10088/32616/A_revised_Felidae_Taxonomy_CatNews.pdf

[5] Meachen J, Schmidt-Küntzel A, Haefele H, Steenkamp G, Robinson JM, Randau M, et al. Cheetah specialization: Physiology and morphology. In: Marker L, Boast LK, Schmidt-Küntzel A, editors. Biodiversity of the World—Cheetahs: Biology and Conservation. 1st ed. San Diego: Elsevier; 2018. pp. 93-105

[6] Hildebrand M. Motions of the running cheetah and horse. Journal of Mammalogy. 1959;**40**:481-495

[7] Hunter L, Hamman D. Cheetah. Cape Town: Struik Publishers; 2003. 144 p

[8] Wilson RP, Griffiths IW, Mills MGL, Carbone C, Wilson JW, Scantlebury DM. Mass enhances speed but diminishes turn capacity in terrestrial pursuit predators. eLife. 2015;**4**:e06487. DOI: 10.7554/eLife.06487

[9] Durant SM, Mitchell N, Groom R, Pettorelli N, Ipavec A, Jacobson A, et al. The global decline of cheetah and what it means for conservation. Proceedings of the National Academy of Sciences of the United States of America. 2017;**114**:528-533. DOI: 10.1073/pnas.1611122114

[10] O'Brien SJ, Wildt DE, Goldman D, Merril CR, Bush M. The cheetah is depauperate in genetic variation. Science. 1983;**221**:459-462. DOI: 10.1126/science.221.4609.459

[11] O'Brien SJ, Roelke ME, Marker L, Newman A, Winkler CA, Meltzer D, et al. Genetic basis for species vulnerability in the cheetah. Science. 1985;**227**:1428-1434. DOI: 10.1126/science.2983425

[12] Schmidt-Küntzel A, Dalton DL, Menotti-Raymond M, Fabiano E, Charruau P, Johnson WE, et al. Conservation genetics of the cheetah: Genetic history and implications for conservation. In: Marker L, Boast LK, Schmidt-Küntzel A, editors. Biodiversity of the World—Cheetahs: Biology and Conservation. 1st ed. San Diego: Elsevier; 2018. pp. 72-89

[13] Caro TM. Cheetahs of the Serengeti Plains: Group Living in an Asocial Species. Chicago: University of Chicago Press; 1994. p. 478

[14] Marker L, Cristescu B, Dickman A, Nghikembua MT, Boast LK, Morrison T, et al. Ecology of free-ranging cheetahs. In: Marker L, Boast LK, Schmidt-Küntzel A, editors. Biodiversity of the World—Cheetahs: Biology and Conservation. 1st ed. San Diego: Elsevier; 2018. pp. 107-116

[15] Durant SM. Living with the enemy: Avoidance of hyenas and lions by cheetahs in the Serengeti. Behavioral Ecology. 2000;**11**:624-632. DOI: 10.1093/beheco/11.6.624

[16] Broekhuis F, Cozzi G, Valeix M, McNutt JW, Macdonald DW. Risk avoidance in sympatric large carnivores: Reactive or predictive? The Journal of Animal Ecology. 2013;**82**:1098-1105. DOI: 10.1111/1365-2656.12077

[17] Mills MGL. Living near the edge: A review of the ecological relationships between large carnivores in the arid Kalahari. African Journal of Wildlife Research. 2015;**45**:127-137. DOI: 10.3957/056.045.0127

[18] Marker L, Dickman AJ, Jeo RM, Mills MGL, Macdonald DW. Demography of the Namibian cheetah. Biological Conservation. 2003;**114**(3):413-425. DOI: 10.1016/S0006-3207(03)00069-7

[19] Dickman A, Rust NA, Boast LK, Wykstra M, Richmond-Coggan L, Klein R, et al. The costs and causes of human-cheetah conflict on livestock and game farms. In: Marker L, Boast LK, Schmidt-Küntzel A, editors. Biodiversity of the World—Cheetahs: Biology and Conservation. 1st ed. San Diego: Elsevier; 2018. pp. 173-186

[20] IUCN/SSC. Regional Conservation Strategy for the Cheetah and African Wild Dog in Eastern Africa. Gland, Switzerland: IUCN/SSC; 2007

[21] IUCN/SSC. Regional Conservation Strategy for the Cheetah and African Wild Dog in Southern Africa. Gland, Switzerland: IUCN/SSC; 2007

[22] IUCN/SSC. Regional Conservation Strategy for the Cheetah and African Wild Dog in Western, Central and Northern Africa. Gland, Switzerland: IUCN/SSC; 2012

[23] RWCP and IUCN/SSC. Regional Conservation Strategy for the Cheetah and African Wild Dog in Southern Africa. Gland, Switzerland: IUCN/SSC; August 2015 (revised and updated)

[24] Marker L, Grisham J, Brewer B. A brief history of cheetah conservation. In: Marker L, Boast LK, Schmidt-Küntzel A, editors. Biodiversity of the World—Cheetahs: Biology and Conservation. 1st ed. San Diego: Elsevier; 2018. pp. 3-14

[25] Jeo RM, Schmidt-Küntzel A, Ballou JD, Sanjayan M. Drivers of habitat loss and fragmentation: Implications for the design of landscape linkages for cheetahs. In: Marker L, Boast LK, Schmidt-Küntzel A, editors. Biodiversity of the World—Cheetahs: Biology and Conservation. 1st ed. San Diego: Elsevier; 2018. pp. 137-149

[26] Marker L, Boast LK, Schmidt-Küntzel A. What does the future hold? In: Marker L, Boast LK, Schmidt-Küntzel A, editors. Biodiversity of the World—Cheetahs: Biology and Conservation. 1st ed. San Diego: Elsevier; 2018. pp. 549-556

[27] Cherin M, Iurino DA, Sardella R, Rook L. *Acinonyx pardinensis* (Carnivora, Felidae) from the early Pleistocene of Pantalla (Italy): Predatory behavior and ecological role of the giant Plio-Pleistocene cheetah. Quaternary Science Reviews. 2014;**87**:82-97. DOI: 10.1016/j.quascirev.2014.01.004

[28] Marker L, Cristescu B, Morrison T, Flyman MV, Horgan J, Sogbohossou EA, et al. Cheetah rangewide status and distribution. In: Marker L, Boast LK, Schmidt-Küntzel A, editors. Biodiversity of the World—Cheetahs: Biology and Conservation. 1st ed. San Diego: Elsevier; 2018. pp. 34-51

[29] Farhadinia MS, Hunter LTB, Jourabchian AR, Hosseini-Zavarei F, Akbari H, Ziaie H, et al. The critically

endangered Asiatic cheetah *Acinonyx jubatus venaticus* in Iran: A review of recent distribution, and conservation status. Biodiversity and Conservation. 2017;**26**:1027. DOI: 10.1007/s10531-017-1298-8

[30] Marker LL, Dickman AJ, Mills MGL, Jeo RM, Macdonald DW. Spatial ecology of cheetahs on north-central Namibian farmlands. Journal of Zoology. 2007;**274**:226-238. DOI: 10.1111/j.1469-7998.2007.00375.x

[31] Belbachir F, Pettorelli N, Wacher T, Belbachir-Bazi A, Durant SM. Monitoring rarity: The critically endangered Saharan cheetah as a flagship species for a threatened ecosystem. PLoS One. 2015;**10**:e0115136. DOI: 10.1371/journal.pone.0115136

[32] Farhadinia MS, Akbari H, Eslami M, Adibi MA. A review of ecology and conservation status of Asiatic cheetah in Iran. Cat News Special Issue [Internet]. 2016 November;**10**:18-26. Available from: https://www.researchgate.net/publication/310794931_A_review_of_ecology_and_conservation_status_of_Asiatic_cheetah_in_Iran

[33] Marker LL. Aspects of cheetah (*Acinonyx jubatus*) biology, ecology and conservation strategies on Namibian farmlands [thesis]. Oxford: University of Oxford; 2002

[34] Durant S. Range-wide conservation planning for cheetah and wild dog. Cat News [Internet]. 2007;**46**:13. Available from: http://www.catsg.org/cheetah/05_library/5_3_publications/D/Durant_2007_Range-wide_Conservation_Planning_for_Cheetah_and_Wild_Dog.pdf

[35] Durant S, Mitchell N, Ipavec A, Groom R. Acinonyx jubatus. In: The IUCN Red List of Threatened Species [Internet]. 2015. p. e.T219A50649567. Available from: DOI: 10.2305/IUCN.UK.2015-4.RLTS.T219A50649567.en

[36] Durant SM, Mitchell N, Groom R, Ipavec A, Woodroffe R, Breitenmoser C, et al. The conservation status of the cheetah. In: Marker L, Boast LK, Schmidt-Küntzel A, editors. Biodiversity of the World—Cheetahs: Biology and Conservation. 1st ed. San Diego: Elsevier; 2018. pp. 533-548

[37] Smithers RHN. Family felidae. In: Meester J, Sezter HW, editors. The Mammals of Africa: An Identification Manual. Washington: Smithsonian Institution Press; 1977. p. 483

[38] Nowell K, Jackson P. Wild cats: Status survey and conservation action plan. In: IUCN/SSC Action Plans or the Conservation of Biological Diversity [Internet]. Vol. 1. 1996. p. 382. Available from: https://portals.iucn.org/library/node/6998

[39] Wachter B, Broekhuis F, Melzheimer J, Horgan J, Chelysheva EV, Marker L, et al. Behavior and communication of free-ranging cheetahs. In: Marker L, Boast LK, Schmidt-Küntzel A, editors. Biodiversity of the World—Cheetahs: Biology and Conservation. 1st ed. San Diego: Elsevier; 2018. pp. 121-133

[40] Bissett C, Parker DM, Bernard RTF, Perry TW. Management-induced niche shift? The activity of cheetahs in the presence of lions. African Journal of Wildlife Research. 2015;**45**:197-203. DOI: 10.3957/056.045.0197

[41] Myers N. The cheetah *Acinonyx jubatus* in Africa. In: IUCN Monograph No. 4. Switzerland: IUCN, Morges; 1975

[42] Kelly MJ, Laurenson MK, Fitzgibbon CD, Collins DA, Durant SM, Frame GW, et al. Demography of the Serengeti cheetah (*Acinonyx jubatus*) population: The first 25 years. Journal of Zoology. 1998;**244**:473-488. DOI: 10.1111/j.1469-7998.1998.tb00053.x

[43] Laurenson MK, Wielebnowski N, Caro TM. Extrinsic factors and juvenile

mortality in cheetahs. Conservation Biology. 1995;**9**:1329-1331. DOI: 10.1046/j.1523-1739.1995.9051327.x-i1

[44] Marker LL, Wilkerson AJP, Sarno RJ, Martenson J, Breitenmoser-Würsten C, O'Brien SJ, et al. Molecular genetic insights on cheetah (*Acinonyx jubatus*) ecology and conservation in Namibia. The Journal of Heredity. 2008;**99**(1): 2-13. DOI: 10.1093/jhered/esm081

[45] Marker L, Rabeil T, Comizzoli P, Clements H, Nghikembua MT, Hayward MW, et al. The status of key prey species and the consequences of prey loss for cheetah conservation in North and West Africa. In: Marker L, Boast LK, Schmidt-Küntzel A, editors. Biodiversity of the World—Cheetahs: Biology and Conservation. 1st ed. San Diego: Elsevier; 2018. pp. 151-159

[46] Fabiano E, Boast LK, Fuller AK, Sutherland C. The use of remote camera trapping to study cheetahs: Past reflections and future directions. In: Marker L, Boast LK, Schmidt-Küntzel A, editors. Biodiversity of the World—Cheetahs: Biology and Conservation. 1st ed. San Diego: Elsevier; 2018. pp. 415-425

[47] Nghikembua M, Harris J, Tregenza T, Marker L. Interactions between bush encroachment and large carnivore habitat selection: A case study on GPS satellite collared cheetahs in northern Namibia. Open Journal of Forestry. 2016;**6**(4):269-280. DOI: 10.4236/ojf.2016.64022

[48] Muntifering JR, Dickman AJ, Perlow LM, Hruska T, Ryan PG, Marker LL, et al. Managing the matrix for large carnivores: A novel approach and perspective from cheetah (*Acinonyx jubatus*) habitat suitability modelling. Animal Conservation. 2006;**9**:103-112. DOI: 10.1111/j.1469-1795.2005.00008.x

[49] Marker LL, Dickman AJ, Mills MGL, Macdonald DW. Cheetahs and ranches in Namibia: A case study. In: Macdonald DW, Loveridge J, editors. Biology, Conservation of Wild Felids. Oxford: Oxford University Press; 2010. pp. 353-372

[50] Nghikembua MT, Lehner F, Ottichilo W, Marker L, Amstrup SC. The impact of climate change on the conservation and survival of the cheetah. In: Marker L, Boast LK, Schmidt-Küntzel A, editors. Biodiversity of the World—Cheetahs: Biology and Conservation. 1st ed. San Diego: Elsevier; 2018. pp. 163-169

[51] Nowell K, Rosen T. Global cheetah conservation policy: A review of international law and enforcement. In: Marker L, Boast LK, Schmidt-Küntzel A, editors. Biodiversity of the World—Cheetahs: Biology and Conservation. 1st ed. San Diego: Elsevier; 2018. pp. 291-303

[52] Tricorache P, Nowell K, Wirth G, Mitchell N, Boast LK, Marker L. Pets and pelts: Understanding and combating poaching and trafficking in cheetahs. In: Marker L, Boast LK, Schmidt-Küntzel A, editors. Biodiversity of the World—Cheetahs: Biology and Conservation. 1st ed. San Diego: Elsevier; 2018. pp. 191-205

[53] CITES. Quotas for trade in specimens of cheetah. In: Eighth Meeting of the Convention of International Trade in Endangered Species of Wild Fauna and Flora. 1992. pp. 1-5

[54] Cristescu B, Schmidt-Küntzel A, Schwartz KR, Traeholt C, Marker L, Fabiano E, et al. A review of population viability analysis and its use in cheetah conservation. In: Marker L, Boast LK, Schmidt-Küntzel A, editors. Biodiversity of the World—Cheetahs: Biology and Conservation. 1st ed. San Diego: Elsevier; 2018. pp. 517-529

[55] Berry H, Bush M, Davidson B, Forge O, Fox B, Grisham J, et al.,

editors. Population and habitat viability assessment for the Namibian cheetah (*Acinonyx jubatus*) and lion (*Panthera leo*). In: Workshop Report. Apple Valley, MN: IUCN/SSC Conservation Breeding Specialist Group; 1997

[56] Munson L, Terio K, Worley M, Jago M, Bagot-Smith A, Marker L. Extrinsic factors significantly affect patterns of disease in free-ranging and captive cheetah (*Acinonyx jubatus*) populations. Journal of Wildlife Diseases. 2005;**41**(3):542-548. DOI: 10.7589/0090-3558-41.3.542

[57] Terio KA, Mitchell E, Walzer C, Schmidt-Küntzel A, Marker L, Citino S. Diseases impacting captive and free-ranging cheetahs. In: Marker L, Boast LK, Schmidt-Küntzel A, editors. Biodiversity of the World—Cheetahs: Biology and Conservation. 1st ed. San Diego: Elsevier; 2018. pp. 349-360

[58] Marker L. A Future for Cheetahs. Cheetah Conservation Fund. Virginia: Cheetah Conservation Fund; 2014. p. 208

[59] Burney DA. The effects of human activities on cheetah (*Acinonyx jubatus*) in the Mara region of Kenya [thesis]. Narirobi: University of Nairobi; 1980

[60] Hughes C, Horgan J, Klein R, Marker L. Cheetah conservation and educational programs. Book chapter. In: Marker L, Boast LK, Schmidt-Küntzel A, editors. Biodiversity of the World—Cheetahs: Biology and Conservation. 1st ed. San Diego: Elsevier; 2018. pp. 251-262

[61] Mwanzia JM, Kock R, Wambua J, Kock N, Jarret O. An outbreak of Sarcoptic mange in the free-living cheetah (*Acinonyx jubatus*) in the Mara region of Kenya. In: Proceedings of American Association of Zoo Veterinarians and American Association of Wildlife Veterinarians Joint Conference. 1995. pp. 105-112

[62] Wykstra M, Combes G, Oguge N, Klein R, Boast LK, Mosimane AW, et al. Improved and alternative livelihoods: The link between poverty alleviation and biodiversity conservation. In: Marker L, Boast LK, Schmidt-Küntzel A, editors. Biodiversity of the World—Cheetahs: Biology and Conservation. 1st ed. San Diego: Elsevier; 2018. pp. 223-236

[63] Jones B, Weaver C. CBNRM in Namibia: Growth, trends, lessons and constraints. In: Suich H, Child B, Spenceley A, editors. Evolution, Innovation in Wildlife Conservation: Parks, Game Ranches to Transfrontier Conservation Areas. London: Earthscan; 2009. pp. 223-242

[64] Powell LA, Kharuxab R, Marker L, Nghikembua MT, Omusula S, Reid RS, et al. Coordination of large landscapes for cheetah conservation. In: Marker L, Boast LK, Schmidt-Küntzel A, editors. Biodiversity of the World—Cheetahs: Biology and Conservation. 1st ed. San Diego: Elsevier; 2018. pp. 239-248

[65] Cristescu B, Lindsey P, Maes O, Bissett C, Mills G, Marker L. Protected areas for cheetah conservation. In: Marker L, Boast LK, Schmidt-Küntzel A, editors. Biodiversity of the World—Cheetahs: Biology and Conservation. 1st ed. San Diego: Elsevier; 2018. pp. 265-272

[66] Marker L, Boast L. Human wildlife conflict 10 years later: Lessons learnt and their application to cheetah conservation. Human Dimensions of Wildlife. 2015;**20**(4):1-8. DOI: 10.1080/10871209.2015.1004144

[67] Marker LL, Dickman AJ, Macdonald DW. Perceived effectiveness of livestock guarding dogs placed on Namibian farms. Rangeland Ecology & Management. 2005;**58**:329-336. DOI: 10.2111/1551-5028(2005)058[0329:PEOLDP]2.0.CO;2

[68] Dickman A, Potgieter G, Horgan J, Stoner K, Klein R, McManus J, et al.

Use of livestock guarding dogs to reduce human-cheetah conflict. In: Marker L, Boast LK, Schmidt-Küntzel A, editors. Biodiversity of the World—Cheetahs: Biology and Conservation. 1st ed. San Diego: Elsevier; 2018. pp. 209-219

[69] Coppinger R, Lorenz J, Glendinning J, Pinardi P. Attentiveness of guarding dogs for reducing predation on domestic sheep. Journal of Range Management. 1983;**36**:275-279. DOI: 10.2307/3898468

[70] Andelt WF. Effectiveness of livestock guarding dogs for reducing predation on domestic sheep. Wildlife Society Bulletin. 1992;**20**:55-62 Retrieved from: www.jstor.org/stable/3782760

[71] Rigg R. Livestock guarding dogs: Their current use world wide. In: IUCN/SSC Canid Specialist Group, Occasional Paper No. 1. 2001

[72] Potgieter GC, Kerley GIH, Marker LL. More bark than bite? The role of livestock guarding dogs in predator control on Namibian farmlands. Oryx. 2016;**50**(3):514-522. DOI: 10.1017/S0030605315000113

[73] Horgan JE. Testing the effectiveness and cost-efficiency of livestock guarding dogs in Botswana [thesis]. Grahamstown: Rhodes University; 2015

[74] Bell D. Cheetah Country Beef Business Plan. Windhoek, Namibia: MeatCo; 2006

Chapter 5

Wild Fodder Yielding Plants in the Protected Areas of Bangladesh

Md Akhter Hossain, Rajib Mahmud, Nikhil Chakma and Mohammed Kamal Hossain

Abstract

Wildlife habitat is degrading worldwide due to myriads of biotic and abiotic reasons. The governments across the world countries are trying to halt the degradation through declaring protected areas (PAs) with potential wildlife habitats and strengthening conservation initiatives. These measures are expected to uphold the richness and diversity of the fodder yielding plants. But there is a dire necessity of information on composition and overall status of the fodder yielding plants for continuous monitoring of these habitats. Moreover, the potentiality of the protected areas can also be judged based on the composition and richness of fodder yielding plants. Having all these in mind, we assessed the composition and conservation status of the fodder yielding plants of all habit forms from three recognized protected areas named Chunati Wildlife Sanctuary, Dudhpukuria-Dhopachari Wildlife Sanctuary, and Madhupur National Park. The study indicated the presence of 306 fodder yielding plant species of all habit forms in the three studied protected areas. This chapter describes the composition, status, habit forms, and nature of occurrences of the wild fodder yielding plants which is expected to be highly helpful in wildlife habitat monitoring and undertaking specific measures for multiplication and conservation of fodder yielding plants.

Keywords: wildlife habitat, national park, forests, habitat monitoring, wildlife sanctuary

1. Introduction

Grasses, shrubs, and different palatable parts of trees, i.e., leaves, flowers, fruits, and seeds, that have nutritive values constitute fodder of wild herbivores [1]. The wildlife population greatly depends on the habitat richness with food, nesting, and breeding environment. The wild fruit and fodder-producing plants play a great role in maintaining ecosystem food supply. The plants are from different habit forms and taxonomic families. Plants from Poaceae, Cyperaceae, Fabaceae, Moraceae, Myrtaceae, and Zingiberaceae families particularly the leguminous plants dominate the fodder-producing plants [2]. There are significant seasonal variation of fodder availability and composition to which the wild animals' nutrition needs are naturally adapted [3]. Insects, birds, chordates, and reptiles have different nesting and breeding natures which varies widely from each other. The habitat resources and overall conditions including food, water, shade, nesting, etc. are needed by a species for survival and reproductive success [4]. Moreover, habitat is organism-specific;

the appropriate combination of necessary abiotic and biotic components for successful reproduction and survival varies by species [5].

The global forests are drastically shrinking day by day due to a huge pressure on forests for conversion to other land use, human settlement, excessive resource extraction, etc. All these are affecting biodiversity negatively particularly the wildlife habitat which is degrading in an alarming way which leads to rapid shrinking of wildlife population and making them threatened. Declaring protected area (PA) is a worldwide strategy accepted for wildlife habitat conservation and ensuring undisturbed breeding ground by halting further fragmentation and degradation of habitat. This chapter presents the status of the protected areas from different corners of Bangladesh in terms of fodder yielding plant composition as well as the importance of fodder plant management for maintaining a healthy wildlife habitat. We identified three protected areas of characteristic features for studying the wild fodder yielding plants so that these represent all the PA of Bangladesh. This chapter also presents a brief account on the protected area management in Bangladesh as well as potentiality of those areas as wildlife habitats.

2. Rationale of the study

Quality of the wildlife habitat largely depends on the nature and composition of vegetation as it determines the nesting, breeding, and feeding potentiality of the habitats. The habitat degradation is causing loss of regeneration of many of the fodder plants. Moreover, overexploitation is also responsible for sharp reduction of the population size of the fodder plants [6]. It is important to detect the food habit of the herbivores and identify the fodder plants and their interactions with animals for sustainable management of the protected areas through wildlife conservation and undisturbed breeding ground [7]. But, in Bangladesh, there is a great dearth of information and research on the wild fodder-bearing trees. Information on the status and availability of plants will be helpful for better planning and management interventions of the PAs as wildlife habitats. The plants used as fodder by the animals are rich in necessary nutrition, i.e., protein, carbohydrate, fiber, etc. The ability of fodder plants to provide this range of nutrients is considered to evaluate potentiality of their nutritive values. Considering the mentioned situations, the study was undertaken to assess the overall composition of fodder plants as well as highlight their contributions for better maintaining a wildlife habitat. However, there is still a need for investigation of the nutritive values as many of the species were not explored yet [6]. We believe this study will fill up the knowledge gap on fodder yielding plant composition in the PA of Bangladesh as well as it will show the way for further research and interventions to habitat conservation and sustainable management.

3. Status of wildlife in Bangladesh

Bangladesh is the home of more than 3883 species of lower faunal groups along with 653 fish species, 49 amphibians, 154 reptiles, 706 birds, and 128 mammals. The fish communities including both freshwater and marine species are so diverse that they account an astonishing 3% of the world's total fish species. In addition to the 383 resident birds, there are 323 migratory birds which visit our country especially during the winter. Both of these represent an amazing 7.2% of the world's total bird species. Mammals constitute 2.28% of the world's mammal species among which seven are marine in nature [8]. Though our wildlife diversity is very rich, but over

DOI: http://dx.doi.org/10.5772/intechopen.85205

Vertebrate groups	No. of species	References
Fish	653	[9]
Amphibians	64	[8]
Reptiles	174	[8]
Birds	711	[8]
Mammals	133	[8]

Table 1.
Number of recorded fauna of Bangladesh.

the last century unfortunately 13 species have become extinct from Bangladesh. Due to continuity of habitat degradation, many more are on the brink of extinction. Different reports indicated that 23% of vertebrates of Bangladesh are facing different levels of threats which are increasing exponentially with rapid habitat destruction. The situation is even more grim for the 57% of reptiles and 36% of mammals which are facing different levels of threats in our country [9]. Recently, the IUCN listed 40 species of mammals, 41 aves, 58 reptiles, and 8 amphibians that are struggling under various degrees of threat of extinction. It is obvious that the present wildlife population is confined and distributed irregularly in limited forest patches of Bangladesh (**Table 1**).

4. Wildlife habitats in Bangladesh

Among the 5 global ecological domains and 20 global ecological zones of the world, 33% of Bangladesh belongs to tropical rain forest GEZ and 67% to tropical moist deciduous forest GEZ of the tropical domains of global ecological domains [10]. The variation in climatic features, i.e., temperature, rainfall, soil, and hydrology, led to the development of 25 bioecological zones with distinct characteristics. Bangladesh has 1.45 million ha of forest land (9.8% of total area) of which 1.21 million ha (84% of forest) is natural forest and 0.24 million ha (16% of forest) is plantations [10].

Vegetation characteristics divided the natural forests of Bangladesh into evergreen/semievergreen, deciduous, and mangrove forest. Noncontinuous freshwater swamp is distributed in the northeast basin. Tropical evergreen and semievergreen forests constituting 44% of natural forest are extended over Chittagong, Chittagong Hill Tracts (CHT), Cox's Bazar, and Sylhet covering an area of 6700 km^2 which is about 4.54% of total landmass of Bangladesh. Dominant native plant species include species of *Dipterocarpus* spp., *Artocarpus* spp., *Ficus* spp., *Syzygium* spp., *Mangifera* spp., *Tectona grandis*, etc. The moist deciduous Sal forest of Bangladesh is mainly consisted of Madhupur tract which is located in the central part covering an area of 340 km^2 [10]. Dominant trees of this forest are *Shorea robusta*, *Lagerstroemia speciosa*, *Dillenia pentagyna*, *Adina cordifolia*, *Terminalia* spp., *Albizia* spp., etc. The Sundarbans, the largest single patch of mangrove forest, is located at the southern extremity of the Padma (Ganges) and Jamuna (Brahmaputra) delta which covers about 5770 km^2 area [11]. Fairly dense evergreen plant species of 10–15 m height is the main feature of this forest. These species are adapted for living under saline condition and regular inundation by the tides. Succulent leaves, stilt roots, pneumatophores, and viviparous germination are the key features of these plants. *Heritiera fomes*, *Excoecaria agallocha*, *Nypa fruticans*, *Sonneratia apetala*, *Rhizophora* spp., *Ceriops decandra*, *Phoenix paludosa*, and *Acrostichum aureum* are the common plants of the Sundarbans. Wetlands of Bangladesh also support a large number of plants and wild animals of the country. Nearly 50% (8 million ha) of the total land surface of the country including river, natural lakes, tanks, reservoir, mangrove forests, estuarine, and seasonally inundated floodplains are considered as wetland.

5. Protected areas of Bangladesh

Bangladesh is situated in the northeastern part of the South Asia region, lying between 20°25′ and 26°38′ north latitude and 88°01′ and 92°40′ east longitude. The country is bordered by India to the north, northeast, and west, Myanmar to the southeast, and the Bay of Bengal covering the southern part with a coastline of 710 km. The climate of Bangladesh is tropical, with maximum summer temperature rising between 32 and 38°C. Annual rainfall ranges between 200 and 400 cm. Biogeographically the country lies at the junction of the Indian and Malayan subregions of the Indo-Malayan Realm and is located very near to the western side of Sino-Japanese region.

5.1 Protected area management strategies

Protected areas are "areas especially dedicated to the protection and maintenance of biological diversity and associated cultural resources, which are managed through legal or other effective means, designated or regulated and managed to achieve specific conservation objectives" [12, 13]. PAs have long been considered as the cornerstone of all national and regional conservation strategies. While it is often argued that they are the most effective and widespread measure for conserving forests and biodiversity [13, 14], the importance of complementary off-reserve management has also been acknowledged [15–17]. Globally, the number of PAs has increased significantly over the last few decades in recognition of their importance for conservation. At present, there are more than 100,000 protected area sites worldwide, covering nearly 12% of the world's land surface [18–20]. Currently there are 34 forest PAs in Bangladesh which represent about 17.5% of the total forest land of the country and 1.8% of country's total land [21].

Bangladesh Forest Department shifts its paradigm of conventional forest management to community-led management for ensuring effective governance of natural resources. A total of 34 reserved forests have been declared as protected area where 23 PAs are managed through active participation of community people which is known as co-management. Community people come forward along with the forest department regarding conservation through ensuring sustainable natural resource management.

5.2 Legal frameworks of the protected areas of Bangladesh

The forestry sector of Bangladesh plays an important role in combating poverty for the people living in and around the forest. The history of forest management in Bangladesh is quite old and was shaped and influenced by colonial forest policy. The Forest Policy, 1979, was the first of its kind and was very much influenced by the colonial policy of forest management [22]. Over time this policy proved ineffective due to various socioeconomic factors such as population growth, poverty, overexploitation of resources, and top-down, centralized management approaches. It was felt by experts, communities, and policy makers that a new dimension to the existing forest policy was needed. The Forest Policy, 1994, specifically recognized the importance of peoples' participation in forest management [23]. Another notable achievement of the 1994 policy was that it has succeeded in bringing tree plantation activities outside the forest area [24].

DOI: http://dx.doi.org/10.5772/intechopen.85205

Most importantly, significant developments in Bangladesh forest legal and policy frameworks took place after the formulation of the 1994 policy [25]. Community participation in the forestry sector of Bangladesh has a long history that can be traced back as early as 1871, to teak plantations of Chittagong Hill Tracts managed by the tribal farmers. However, participation of community people in the forestry sector officially began in the 1980s. Donor-assisted community forestry project was the first attempt of its kind in the northwestern districts of Bangladesh. It gradually spread to other parts of the country through various projects and forms such as the Thana Afforestation and Nursery Development Project (TANDP), the Coastal Greenbelt Project (CGP), and the Forestry Sector Project (FSP). Despite the initial success in achieving physical targets, i.e., increase of plantation coverage, these projects failed to develop a mechanism to attract and engage local communities. They lacked institutional, personal, and community capacity building, legitimacy on usufruct rights, active community participation, and devolution of the decision-making power under the continued influence of "command and control" strategies. The introduction of co-management in the forest PAs is an effort to overcome these limitations to incorporate active community participation as a core aspect of PA governance [23]. The government of Bangladesh started introducing and implementing co-management in five forest PAs under a pilot project titled Nishorgo Support Project (hereafter referred to as NSP) for a period of 5 years (2004–2009) [26]. Many countries have already developed enabling legal and policy frameworks to support community rights and access and have thereby offered better incentives in the governance of the PAs and the resources sustained by them.

5.3 Protected areas as a potential wildlife habitat

As an effective tool, the protected areas are recognized internationally for the conservation of biodiversity. Currently the PAs of Bangladesh represent most of the ecosystems and thus include all habitats and species that are vital for conservation. The Bangladesh Forest Department under the Ministry of Environment, Forest and Climate Change manages a network of 17 national parks, 20 wildlife sanctuaries, 2 special biodiversity conservation areas, 1 marine protected area, 2 vulture safe zones, 2 botanical gardens, 2 safari parks, 10 eco-parks, and 1 aviary park. The total area under this protected area status covers 618253.49 ha of forest land and represents 4.19% area of the country [27]. The primary purpose of these sites is to conserve and protect habitat for wildlife, including migratory birds, species at risk, and other species of national interest.

5.4 Fodder plants in the protected areas of Bangladesh

As a sample of protected areas of Bangladesh, we reviewed the floristic studies [28–30] conducted in three protected areas of Bangladesh with characteristic features. They are "Chunati Wildlife Sanctuary," Dudhpukuria-Dhopachari Wildlife Sanctuary, and "Madhupur National Park." We used the plant data collected during the field survey as secondary data for assessing the fodder yielding plants with due permission from the respective authors. The identified plants were then explored with their use and conservation status following the encyclopedia of flora and fauna of Bangladesh [31].

Chunati Wildlife Sanctuary, established in 1986, is familiar as the habitat and breeding ground of the Asian elephant (*Elephas maximus*). It is one of the oldest PAs of Bangladesh rich with 691 plants from all habit forms [28]. In addition to the Asian elephant, Chunati harbors 26 species of amphibians, 54 reptiles, 252 birds, and 40 mammals [32]. **Tables 2** and **3** provide a detailed account of the flora and fauna of the

Habit forms	Number of species reported from the selected protected areas					
	Chunati Wildlife Sanctuary		Dudhpukuria-Dhopachari Wildlife Sanctuary		Madhupur National Park	
	Total	Fodder yielding	Total	Fodder yielding	Total	Fodder yielding
Trees	240	81	182	61	139	70
Shrubs	102	17	125	21	48	10
Herbs	211	61	200	70	136	43
Climbers	106	25	71	8	46	22
Ferns	19	4	17	4	9	2
Epiphytes	7	—	7	—	7	—
Parasites	6	—	6	—	—	—
Total	691	188	608	164	385	147

Table 2.
Flora of the selected protected areas [28–30].

Groups	Number of species in the selected protected areas		
	Chunati Wildlife Sanctuary	Dudhpukuria-Dhopachari Wildlife Sanctuary	Madhupur National Park
Amphibians	26	25	17
Birds	252	231	120
Fish	10	23	—
Mammals	40	50	27
Reptiles	54	56	28

Table 3.
Fauna of the selected protected areas [28–30].

selected protected areas. Dudhpukuria-Dhopachari Wildlife Sanctuary is a comparatively new protected area that is declared in 2010. It covers an area of 4716 ha rich in both floral and faunal diversity. The wildlife sanctuary harbors 608 plant species and 385 wildlife [29]. However, the Asian elephant is also the flagship animal of this PA. Madhupur National Park, also one of the oldest protected areas, was declared in 1982. It is situated in the central region of Bangladesh covered with mainly deciduous *Shorea robusta*. It harbors 385 plant species from all habit forms and 192 wildlife including amphibians, birds, mammals, and reptiles (**Tables 2** and **3**).

We identified the wild fodder yielding flora of different habit forms following the encyclopedia of flora and fauna of Bangladesh [31]. The review indicated that each of the protected areas harbors a substantial number of fodder yielding plants from different habit forms (**Table 2**).

5.4.1 Trees

A total of 112 tree species belonging to 71 genera and 32 families were found to yield part of it (i.e., leaves, branch, fruit, seed, flower, etc.) as fodder. A comparison number of species in the selected PAs indicated that CWS has 87 species, whereas DDWS and MNP showed 69 and 67 species, respectively (**Table 4**).

DOI: http://dx.doi.org/10.5772/intechopen.85205

SN	Botanical name	Local name	Family	Conservation status	Density in PAs (stem/ha)		
					CWS	DDWS	MNP
1	*Acacia mangium* Willd.	Mangium	Mimosaceae	LC	24.3	0.4	0.4
2	*Acronychia pedunculata* (L.) Miq.	Bonjamir, Jairgola	Rutaceae	NE	0.2	10.6	
3	*Aegle marmelos* (L.) Corr.	Bel	Rutaceae	LC	0.1	0.4	0.4
4	*Alangium chinense* (Lour.) Harms	Marleza Gachh	Alangiaceae	NE (rare)		3.6	
5	*Albizia chinensis* (Osb.) Merr.	Chakua Koroi	Mimosaceae	LC	4	4.8	0.9
6	*Albizia lebbeck* (L.) Benth. & Hook	Kala Koroi	Mimosaceae	LC			1.3
7	*Albizia odoratissima* (L. f.) Benth.	Tetoyakoroi	Mimosaceae	LC	0.4	4.4	
8	*Anacardium occidentale* L.	Kajubadam	Anacardiaceae	LC	0.8		
9	*Annona squamosa* L.	Ata	Annonaceae	LC	0.4		0.4
10	*Antidesma acidum* Retz.	Elena	Euphorbiaceae	LC	0.8	0.4	
11	*Antidesma acuminatum* Wall. in Wight.	Chokoi	Euphorbiaceae	LC			0.4
12	*Antidesma bunius* (L.) Spreng.	Wishwar choa, Banshial Boka	Euphorbiaceae	LC	1.3	0.4	
13	*Antidesma ghaesembilla* Gaertn.	Chokoi, Elena	Euphorbiaceae	LC	0.2		2.6
14	*Aphanamixis polystachya* (Wall.) R.N. Parker.	Ptiraj	Mimosaceae	LC	0.8	3.8	0.4
15	*Aporosa dioica* (Roxb.) Mull.Arg.	Castoma	Euphorbiaceae	NE	3.8		
16	*Aporosa* sp.	Kharjon	Euphorbiaceae	NE			23.3
17	*Areca catechu* L.	Supari	Arecaceae	LC	0.2		0.4
18	*Artocarpus chama* Buch.-Ham. *ex* Wall.	Chapalish, Chambal	Moraceae	NE (rare)	3.9	17.6	0.4
19	*Artocarpus heterophyllus* Lamk.	Kanthal	Moraceae	NE (rare)	4.3	0.2	0.4
20	*Artocarpus lacucha* Buch.-Ham.	Borta	Moraceae	LC	2.4	4.4	1.3
21	*Averrhoa bilimbi* L.	Belombo	Oxalidaceae	LC	0.2		
22	*Averrhoa carambola* L.	Kamranga	Oxalidaceae	LC	0.2		
23	*Baccaurea ramiflora* Lour.	Lotkon	Euphorbiaceae	LC	0.1	0.2	
24	*Borassus flabellifer* L.	Tal	Arecaceae	LC	0.2		0.4
25	*Bridelia retusa* (L.) A. Juss.	Kata Kushui, Kata Koi	Euphorbiaceae	LC		0.6	

SN	Botanical name	Local name	Family	Conservation status	Density in PAs (stem/ha)		
					CWS	DDWS	MNP
26	*Buchanania lancifolia* Roxb.		Anacardiaceae	NE (rare)	0.2		
27	*Callicarpa arborea* Roxb.	Bormala, Khoja	Verbenaceae	LC	7.4	6.8	0.4
28	*Calophyllum polyanthum* Wall. *ex* Choisy	Chandua, Kamdeb	Clusiaceae	NE (rare)		0.6	
29	*Cassia fistula* L.	Sonalu	Caesalpiniaceae	LC	0.5	1	1.7
30	*Citrus reticulata* Blanco	Komla	Rutaceae	LC	0.1		0.4
31	*Clausena heptaphylla* (Roxb.) Wight & Arn. *ex* Steud.	Karan phal, Panbahar, sada Moricha	Rutaceae	LC	1.12	0.2	
32	*Cleistocalyx nervosum* (DC.) Kosterm. var. paniala (Roxb.) J. Parn. & P. Chantaranothai		Myrtaceae	LC	0.2		2.2
33	*Cocos nucifera* L.	Narikel	Arecaceae	LC	0.2	0.6	0.4
34	*Cordia dichotoma* Forst. f.	Bolla gota, Bohal, Bhola	Boraginaceae	LC			0.4
35	*Cordia dichotoma* Forst. f.	Bohal	Boraginaceae	LC	0.3		0.4
36	*Crateva magna* (Lour.) DC.		Capparaceae	LC	0.3		
37	*Cryptocarya amygdalina* Nees.	Ojha	Lauraceae	NE (rare)	2.5	3.4	21.9
38	*Dalbergia sissoo* Roxb.	Sissoo	Fabaceae	LC	0.1		
39	*Dillenia indica* L.	Chalta	Dilleniaceae	LC	0.2	0.2	
40	*Dillenia scabrella* Roxb. *ex* Wall.	Ajuli, Ajugi	Dilleniaceae	LC	4.1	5.8	0.4
41	*Diospyros blancoi* A. DC.	Bilati gab	Ebenaceae	LC	0.2		0.4
42	*Diospyros malabarica* (Desr.) Kostel.	Deshi gab	Ebenaceae	LC	0.2	1.4	
43	*Elaeis guineensis* Jacq.	Palm oil	Arecaceae	NE	0.3		0.4
44	*Elaeocarpus floribundus* Blume	Titpai	Elaeocarpaceae	LC	0.2	1.8	1.3
45	*Elaeocarpus tectorius* (Lour.) Poir.	Jalpai	Elaeocarpaceae	LC	2.5	2.2	
46	*Ficus auriculata* Lour.	Lal Dumur	Moraceae	LC	1.6	0.8	
47	*Ficus benghalensis* L.	Bot	Moraceae	LC	0.6	1.2	1.7
48	*Ficus hispida* L. f.	Dumur	Moraceae	LC	26.9	4.6	0.9
49	*Ficus lanceolata* Buch.-Ham. *ex* Roxb.		Moraceae	V	0.3		

SN	Botanical name	Local name	Family	Conservation status	Density in PAs (stem/ha)		
					CWS	DDWS	MNP
50	*Ficus racemosa* L.	Dumur, Jagyadumur	Moraceae	LC	0.3	2.4	0.9
51	*Ficus religiosa* L.	Bot	Moraceae	LC			0.4
52	*Ficus rumphii* Bl.	Bot	Moraceae	LC			0.4
53	*Ficus semicordata* Buch.-Ham. *ex* Smith	Chokorgola	Moraceae	NE	0.9	0.8	
54	*Ficus tinctoria* G. Forst. subsp. gibbosa (Blume) Corner		Moraceae	NE (rare)	0.1		
55	*Ficus variegata* Blume		Moraceae	NE	0.4	5.2	
56	*Ficus virens* Ait.	Pakur, Pakar, Paikur	Moraceae	LC	0.1		0.9
57	*Firmiana colorata* (Roxb.) R. Br.	Udal	Sterculiaceae	LC	0.2		
58	*Flacourtia jangomas* (Lour.) Raeusch.	Painnagola	Flacourtiaceae	LC	1.2	0.2	0.9
59	*Garcinia cowa* Roxb. *ex* DC.	Cao	Clusiaceae	NE (rare)	9.3	5.2	2.2
60	*Garcinia lanceaefolia* Roxb.		Clusiaceae	NE	0.2		
61	*Garcinia xanthochymus* Hook. f. *ex* T. Anders.	Tamal, Dephal	Clusiaceae	LC		1	
62	*Garuga pinnata* Roxb.	Bhadi, Silbhadi, Jeolbhadi	Burseraceae	LC	0.2	7	3.9
63	*Grewia nervosa* (Lour.) Panigr.	Datoi	Tiliaceae	LC	8.5	19.2	22.4
64	*Grewia sapida* Roxb. *ex* DC.	Naricha	Tiliaceae	LC	0.2		
65	*Grewia tiliifolia* Vahl.	Pholsa, Dhomoni	Tiliaceae	LC		0.6	
66	*Grewia serrulata* DC.	Naricha	Tiliaceae	LC			0.4
67	*Hevea brasiliensis* (Willd. *ex* A. Juss.) Mull.Arg.	Rubber	Euphorbiaceae	LC			0.4
68	*Hydnocarpus laurifolius* (Dennst.) Sleum.	Hiddigach	Flacourtiaceae	NE (rare)	1.4	5.2	
69	*Lannea coromandelica* (Houtt.) Merr.	Jialbhadi	Anacardiaceae	LC	4	0.2	3.9
70	*Lepisanthes rubiginosa* (Roxb.) Leenh.		Sapindaceae	LC	0.2	0.2	
71	*Lepisanthes senegalensis* (Poir.) Leenh.	Gotaharina	Sapindaceae	LC	0.2	0.2	

SN	Botanical name	Local name	Family	Conservation status	Density in PAs (stem/ha)		
					CWS	DDWS	MNP
72	*Litchi chinensis* Sonn.	Litchu, Lychee	Sapindaceae	LC	1.0		0.4
73	*Maesa indica* (Roxb.) A. DC.	Maesa, Moricha, Romjani	Myrsinaceae	CD	0.1	2.2	
74	*Mallotus philippensis* (Lamk.) Mull.Arg.	Sinduri	Euphorbiaceae	CD	0.6	0.2	60. 8
75	*Mangifera indica* L.	Am	Anacardiaceae	LC	2.9	0.4	0.9
76	*Mangifera sylvatica* Roxb.	Uriam	Anacardiaceae	V	0.1	0.2	
77	*Manilkara zapota* (L.) P. van Royen	Sofeda	Sapotaceae	LC			0.4
78	*Miliusa velutina* (Dunal) Hook. f.	Gandhi gajari	Annonaceae	LC			3.0
79	*Moringa oleifera* Lamk.	Sajna	Moringaceae	NE			0.9
80	*Peltophorum pterocarpum* (DC.) K. Heyne	Radhachura, Halud Krisnachura	Caesalpiniaceae	LC	0.4		0.4
81	*Phoebe lanceolata* (Ness) Ness	Chaongri, Dulia	Lauraceae	NE		0.2	
82	*Phoenix acaulis* Roxb.	Bon Khejur, Khudi khejur	Arecaceae	V			0.4
83	*Phoenix sylvestris* Roxb.	Khejur	Arecaceae	LC	0.2	0.4	0.4
84	*Phyllanthus emblica* L.	Amloki	Euphorbiaceae	LC	2.4	2	0.9
85	*Protium serratum* (Wall. *ex* Colebr.) Engl.	Gotgutia	Burseraceae	LC	2.1	12.2	32.8
86	*Psidium guajava* L.	Payara	Myrtaceae	LC	4.2	0.2	0.4
87	*Samanea saman* (Jacq.) Merr.	Raintree	Mimosaceae	LC	0.4		0.9
88	*Sapium baccatum* Roxb.	Cham phata	Euphorbiaceae	LC	0.4	3	
89	*Schleichera oleosa* (Lour.) Oken.	Joyna, Kusum	Sapindaceae	NE			14.2
90	*Semecarpus anacardium* L.f.	Bheula, Bhela	Anacardiaceae	NE			11.2
91	*Senna siamea* (Lamk.) Irwin & Barneby	Minjiri	Caesalpiniaceae	LC	3.6	1.8	0.4
92	*Spondias pinnata* (L.f.) Kurz	Bon-Amra, Piala	Anacardiaceae	LC		3.4	0.4
93	*Sterculia hamiltonii* (O. Kuntze) Adelb.		Sterculiaceae	LC		0.2	
94	*Streblus asper* Lour.	Sheora, Harba	Moraceae	LC	1.5	2	0.9
95	*Syzygium balsameum* (Wight) Walp.	Buti Jam	Myrtaceae	LC		1	

DOI: http://dx.doi.org/10.5772/intechopen.85205

SN	Botanical name	Local name	Family	Conservation status	Density in PAs (stem/ha)		
					CWS	DDWS	MNP
96	*Syzygium claviflorum* (Roxb.) A. M. Cowan & J. M. Cowan		Myrtaceae	LC	3.8	0.4	
97	*Syzygium cumini* (L.) Skeels	Kalojam	Myrtaceae	LC	1.3	0.8	0.9
98	*Syzygium cymosum* DC.	Khudi Jam	Myrtaceae	NE		0.2	
99	*Syzygium firmum* Thw.	Dhaki jam	Myrtaceae	LC	7.5	1.8	0.4
100	*Syzygium fruticosum* (Wall.) Masamune	Putijam	Myrtaceae	LC	13.0	1	6.5
101	*Syzygium jambos* (L.) Alston	Gulapjam	Myrtaceae	LC	0.2		
102	*Syzygium praecox* (Roxb.) Rathakr. & N. C. Nair		Myrtaceae	NE	0.2		
103	*Syzygium tetragonum* Wall. *ex* Kurz.	Pholda jam, Lal Pholda	Myrtaceae	NE		3	
104	*Tamarindus indica* L.	Tentul	Caesalpiniaceae	LC	0.1	0.6	0.9
105	*Terminalia bellirica* (Gaertn.) Roxb.	Bohera	Combretaceae	LC	7.5	10.6	29.3
106	*Terminalia catappa* L.	Katbadam	Combretaceae	LC	0.2		
107	*Terminalia chebula* Retz.	Haritaki	Combretaceae	V	0.8	0.6	4.7
108	*Tetrameles nudiflora* R. Br.	Chandul, Maina Kat	Datiscaceae	NE	0.1	1.8	
109	*Trema orientalis* (L.) Blume	Jiban, Naricha	Ulmaceae	LC	5.1	0.2	0.4
110	*Vitex glabrata* R.Br.	Goda arsol, Hakuni gach	Verbenaceae	LC	1.0	1.4	0.9
111	*Vitex peduncularis* Wall. *ex* Schauer	Goda	Verbenaceae	NE (rare)	3.3	11	0.9
112	*Ziziphus mauritiana* Lamk.	Boroi	Rhamnaceae	LC	0.2		0.4

Table 4.
List of fodder yielding trees in three selected protected areas [here, LC, least concern; NE, not evaluated; NE (rare), not evaluated but seems to be rare].

Density of the fodder yielding tree species varied greatly with PAs. *Ficus hispida* was having the highest stem density in CWS, whereas in DDWS *Grewia nervosa* and *Artocarpus chama* were having the highest density. On the other hand, *Mallotus philippensis* and *Protium serratum* were the two mostly dense tree species in MNP. There were 15 fodder yielding exotic tree species in the three protected areas. The studies indicated that density of very few species was good (10 stems/ha); however most of them are having very poor density which apparently seems not indicative of a rich wildlife habitat.

5.4.2 Shrubs

There were 27 fodder yielding shrubby species recorded from the selected three protected areas. These species taxonomically belong to 23 genera and 15 families (**Table 5**). Both CWS and DDWS were represented by 17 shrubby fodder yielding species, whereas MNP showed 14 species indicating its comparative inferiority of supporting wildlife. However, *Cajanus cajan* and *Manihot esculenta* were the two exotic fodder species recorded from the cultivation sites of MNP and CWS.

SN	Botanical name	Local name	Family	Conservation status	Occurrence in selected PAs		
					CWS	DDWS	MNP
1	*Bambusa tulda* Roxb.	Mitinga, Mitinga, Mirtinga, Taralla, Tolla bansh	Poaceae	LC	✓	✓	✓
2	*Bambusa vulgaris* Schrad. *ex* Wendl.	Baijja, Baria, Jowa Bansh, Bangla Bans, Ora Bansh	Poaceae	LC	✓	✓	✓
3	*Bauhinia acuminata* L.		Caesalpiniaceae	LC	✓		
4	*Bridelia stipularis* (L.) Blume	Sitki (climbing)	Euphorbiaceae	LC	✓	✓	✓
5	*Caesalpinia hymenocarpa* (Prain) Hattink.		Caesalpiniaceae	NE (rare)		✓	
6	*Cajanus cajan* (L.) Millsp.	Arhor, Sarata alu, Sortai alu	Fabaceae	LC	✓		✓
7	*Capparis zeylanica* L.		Capparaceae	LC	✓	✓	
8	*Citrus aurantifolia* (Christm. & Panzer) Swingle	Lebu	Rutaceae	LC	✓	✓	✓
9	*Clausena suffruticosa* (Roxb.) Wight & Arn.	Sadamoricha	Rutaceae	LC	✓		
10	*Clerodendrum serratum* (L.) Moon		Verbenaceae	NE	✓		
11	*Crotalaria spectabilis* Roth	Pipli-jhunjan	Fabaceae	LC		✓	
12	*Dendrocalamus longispathus* (Kurz) Kurz	Ora	Poaceae	NE		✓	
13	*Grewia asiatica* L.	Pholsa	Tiliaceae	NE (rare)		✓	✓
14	*Grewia serrulata* DC.	Panisara, Pichandi, Khulla damor	Tiliaceae	LC		✓	✓
15	*Helicia erratica* Hook. f.		Proteaceae	NT	✓		

DOI: http://dx.doi.org/10.5772/intechopen.85205

SN	Botanical name	Local name	Family	Conservation status	Occurrence in selected PAs		
					CWS	DDWS	MNP
16	*Maclura cochinchinensis* (Lour.) Corner		Moraceae	LC		✓	
17	*Manihot esculenta* Crantz	Cassava, Gach alu	Euphorbiaceae	LC			✓
18	*Melocanna baccifera* (Roxb.) Kurz	Muli	Poaceae	LC	✓	✓	
19	*Murraya koenigii* (L.) Spreng.		Rutaceae	LC	✓		
20	*Phlogacanthus thyrsiformis* Roxb *ex* D. J. Mabberley		Acanthaceae	NE			✓
21	*Premna esculenta* Roxb.	Lalana	Verbenaceae	LC	✓	✓	✓
22	*Punica granatum* L.	Dalim	Punicaceae	LC			✓
23	*Sarcochlamys pulcherrima* Gaudich.	Jangallya shak, Maricha	Urticaceae	NE	✓	✓	
24	*Solanum melongena* L.	Begun	Solanaceae	LC	✓		✓
25	*Solanum torvum* Sw.	Tit begun, Gota begun	Solanaceae	LC	✓	✓	
26	*Ziziphus oenoplia* (L.) Mill.	Bonboroi, Toktoki kanta, Tokni boroi	Rhamnaceae	LC	✓	✓	✓
27	*Ziziphus rugosa* Lamk.	Jangli Boroi, Anoi, Anoi gota, Anari gota	Rhamnaceae	NE		✓	✓

Table 5.
List of fodder yielding shrubs reported from the selected protected areas [here, LC, least concern; NE, not evaluated; NE (rare), not evaluated but seems to be rare].

5.4.3 Herbs

The review revealed a total of a higher number of fodder yielding herbs occurring in the protected areas. One hundred twenty-one herbaceous species belonging to 82 genera and 29 families were recorded from the protected areas (**Table 6**). DDWS was represented with the highest number of herb species (70 species) which was followed by CWS and MNP with 60 and 39 herb species, respectively. A substantial number of (15 species) herbs that are reported growing in and around the protected areas were introduced in Bangladesh at different times, and most of these were found to be cultivated in the adjacent forest areas of the protected areas. Wildlife takes advantages of cultivation by raiding them for food especially during the cultivation season. The conservation status of three fodder yielding herbs was vulnerable, i.e., *Colocasia oresbia*, *Homalomena coerulescens*, and *Polygala furcata*.

SN	Botanical name	Local name	Family	Conservation status	Occurrence in selected PAs		
					CWS	DDWS	MNP
1	*Acroceras tonkinense* (Balansa) C.E. Hubb. *ex* Bor		Poaceae	LC	✓		
2	*Actinoscirpus grossus* (L.f.) Goetgh & D. A. Simpson	Kasuru, Kasari, Kesar	Cyperaceae	LC		✓	
3	*Allium cepa* L.	Piyaj	Liliaceae	LC	✓		
4	*Allium sativum* L.	Rashun	Liliaceae	LC	✓		
5	*Alocasia macrorrhizos* (L.) G. Don	Mankatchu	Araceae	LC			✓
6	*Alternanthera philoxeroides* (Mart.) Griseb.	Helencha, Malancha shak	Amaranthaceae	LC			✓
7	*Amaranthus spinosus* L.	Kantashakh, Kata Notay, Khoira kanta	Amaranthaceae	LC	✓		✓
8	*Amaranthus tricolor* L.	Lalshakh, Danga, Data shak	Amaranthaceae	LC	✓		✓
9	*Amischophacelus axillaris* (L.) Rolla Rao & Kamm.	Baghanulla	Commelinaceae	LC		✓	
10	*Amorphophallus bulbifer* (Roxb.) Blume	Amla-bela, Jongle Ol.	Araceae	LC		✓	✓
11	*Amorphophallus paeoniifolius* (Dennst.) Nicolson	Bag katchu, Batema katchu	Araceae	LC			✓
12	*Ananas comosus* (L.) Merr.	Anarosh	Bromeliaceae	LC	✓		✓
13	*Aponogeton echinatus* Roxb.	Ghechu	Aponogetonaceae	CD		✓	
14	*Aponogeton natans* (L.) Engl. & Krause		Aponogetonaceae	NT		✓	
15	*Arundo donax* L.		Poaceae	LC	✓		
16	*Axonopus compressus* (Sw.) P. Beauv.	Ghora dubo Har, Farak pata	Poaceae	LC	✓	✓	✓
17	*Bothriochloa bladhii* (Retz.) S. T. Blake	Gandha Gourana	Poaceae	LC		✓	
18	*Brachiaria decumbens* Stapf		Poaceae	LC		✓	
19	*Brachiaria distachya* (L.) Stapf	Corighas	Poaceae	LC	✓	✓	✓
20	*Brachiaria kurzii* (Hook. f.) *A. Camus*		Poaceae	LC			✓
21	*Brachiaria reptans* (L.) Gard. & Hubb.		Poaceae	LC	✓		

DOI: http://dx.doi.org/10.5772/intechopen.85205

SN	Botanical name	Local name	Family	Conservation status	Occurrence in selected PAs		
					CWS	DDWS	MNP
22	*Brassica oleracea* L. var. botrytis L.	Phulkopi	Brassicaceae	LC	✓		
23	*Bryophyllum pinnatum* (Lamk.) Oken	Pathorkutchi, Pathorchura	Crassulaceae	LC	✓		✓
24	*Capillipedium assimile* (Steud.) A Camus		Poaceae	DD		✓	
25	*Capsicum annuum* L.	Morich	Solanaceae	LC			✓
26	*Capsicum frutescens* L.	Morich	Solanaceae	LC	✓		
27	*Carica papaya* L.	Papaya, Pape	Caricaceae	LC	✓		✓
28	*Centella asiatica* (L.) Urban	Thankuni	Apiaceae	LC	✓	✓	✓
29	*Chenopodium album* L.	Batua Shakh	Chenopodiaceae	LC	✓		
30	*Chrysopogon aciculatus* (Retz.) Trin	Lengra, Premkanta	Poaceae	LC	✓	✓	✓
31	*Colocasia esculenta* (L.) Schott	Katchu	Araceae	LC	✓	✓	✓
32	*Colocasia fallax* Schott		Araceae	LC		✓	
33	*Colocasia oresbia* A. Hay	Sadakachu	Araceae	V		✓	
34	*Commelina benghalensis* L.	Dholpata, Kanchira	Commelinaceae	LC		✓	✓
35	*Commelina sikkimensis* C.B. Clarke	Batbaithia Shag	Commelinaceae	CD		✓	
36	*Corchorus capsularis* L.	Pat shakh	Tiliaceae	LC			✓
37	*Curcuma longa* L.	Halud	Zingiberaceae	LC	✓		✓
38	*Cyanotis cristata* (L.) D. Don		Commelinaceae	LC		✓	
39	*Cymbopogon citratus* (DC) Stapf	Dhan Sabarang, Lemon Ghas	Poaceae	CD		✓	
40	*Cynodon arcuatus* J. S. Presl *ex* C. B. Presl		Poaceae	LC		✓	
41	*Cynodon dactylon* (L.) Pers.	Durba grass	Poaceae	LC	✓	✓	✓
42	*Cyperus corymbosus* Rottb.	Gola Methi	Cyperaceae	NE		✓	
43	*Cyperus cyperoides* (L.) O. Ktze.	Kucha, Kusha	Cyperaceae	LC		✓	
44	*Cyperus difformis* L.	Behua	Cyperaceae	LC		✓	✓
45	*Cyperus digitatus* Roxb.	Behua	Cyperaceae	LC		✓	
46	*Cyperus distans* L. f.	Pani Malanga	Cyperaceae	LC		✓	
47	*Cyperus laxus* Lamk var. laxus		Cyperaceae	LC		✓	✓

SN	Botanical name	Local name	Family	Conservation status	Occurrence in selected PAs		
					CWS	DDWS	MNP
48	*Cyperus rotundus* L.	Mutha	Cyperaceae	LC		✓	
49	*Cyperus tuberosus* Rottb.		Cyperaceae	LC	✓	✓	
50	*Cyrtococcum oxyphyllum* (Steud.) Stapf		Poaceae	LC		✓	
51	*Cyrtococcum patens* (L.) A. Camus		Poaceae	LC	✓	✓	
52	*Dactyloctenium aegyptium* (L.) P. Beauv.	Makra	Poaceae	LC		✓	
53	*Desmostachya bipinnata* (L.) Stapf		Poaceae	LC	✓		
54	*Dichanthium caricosum* (L.) A. Camus	Detara	Poaceae	LC		✓	
55	*Digitaria bicornis* (Lamk.) Roem. & Schult. *ex* Loud		Poaceae	NE			✓
56	*Digitaria sanguinalis* (L.) Scop.		Poaceae	LC	✓		
57	*Echinochloa crus-galli* (L.) P. Beauv.	Bara Shama-ghas	Poaceae	LC	✓	✓	
58	*Eichhornia crassipes* (Mart.) Solms		Pontederiaceae	LC	✓		
59	*Eleusine indica* (L.) Gaertn.	Malankuri	Poaceae	LC	✓	✓	
60	*Eragrostis ciliaris* (L.) R. Br.		Poaceae	LC	✓		✓
61	*Eragrostis lehmanniana* Nees		Poaceae	NE		✓	
62	*Eragrostis tenella* (L.) P. Beauv. *ex* Roem. & Schult.	Koni Ghas	Poaceae	LC	✓	✓	
63	*Eragrostis unioloides* (Retz.) Nees *ex* Steud.		Poaceae	LC	✓		
64	*Eriochloa procera* (Retz.) C. E. Hubb.		Poaceae	LC	✓		✓
65	*Euphorbia heterophylla* L.		Euphorbiaceae	NE (rare)		✓	
66	*Fuirena umbellata* Rottb.		Cyperaceae	LC		✓	
67	*Homalomena coerulescens* Jungh.		Araceae	V		✓	
68	*Hydrolea zeylanica* (L.) Vahl		Hydrophyllaceae	LC	✓		
69	*Hymenachne pseudointerrupta* C. Muell.		Poaceae	LC	✓		

SN	Botanical name	Local name	Family	Conservation status	Occurrence in selected PAs		
					CWS	DDWS	MNP
70	*Imperata cylindrica* (L.) P. Beauv.	Chhan, Chau, Kash	Poaceae	LC	✓	✓	
71	*Juncus prismatocarpus* R.Br.		Juncaceae	LC		✓	✓
72	*Kyllinga brevifolia* Rottb.		Cyperaceae	LC		✓	
73	*Kyllinga bulbosa* Beauv.		Cyperaceae	LC		✓	
74	*Kyllinga nemoralis* (J. R. Forst. & G. Forst.) Dandy *ex* Hutchins. & Dalziel		Cyperaceae	LC	✓	✓	
75	*Lasia spinosa* (L.) Thw.		Araceae	LC	✓		
76	*Lemna perpusilla* Torrey		Lemnaceae	LC		✓	
77	*Leucas indica* (L.) R.Br. *ex* Vatke	Dandakolas, Haldusha, Sweetadrone	Lamiaceae	LC		✓	
78	*Lophatherum gracile* Brongn.		Poaceae	LC			✓
79	*Lycopersicon esculentum* Mill.	Tomato	Solanaceae	LC	✓		✓
80	*Mollugo pentaphylla* L.		Molluginaceae	LC			✓
81	*Monochoria hastata* (L.) Solms	Baranukha	Pontederiaceae	LC	✓	✓	
82	*Monochoria vaginalis* (Burm. f.) Presl	Nukha, Sarkachu	Pontederiaceae	LC	✓	✓	
83	*Musa ornata* Roxb.	Ramkola	Musaceae	CD	✓	✓	
84	*Musa paradisiaca* L.	Champa kola	Musaceae	LC	✓		✓
85	*Oplismenus burmannii* (Retz.) P. Beauv.		Poaceae	LC		✓	
86	*Oryza sativa* L.	Dhan	Poaceae	CD	✓		✓
87	*Panicum maximum* Jacq.		Poaceae	LC	✓		
88	*Panicum paludosum* Roxb.	Barti, Borali, Kalash Nar	Poaceae	LC		✓	
89	*Panicum repens* L.		Poaceae	LC	✓		
90	*Paspalidium flavidum* (Retz.) A. Camus	Bolai Mandi, Karin Ghas	Poaceae	LC		✓	
91	*Paspalum conjugatum* Bergius		Poaceae	LC		✓	
92	*Paspalum longifolium* Roxb.		Poaceae	NE		✓	
93	*Paspalum orbiculare* G. Forst.		Poaceae	LC		✓	
94	*Paspalum scrobiculatum* L.		Poaceae	LC	✓		

SN	Botanical name	Local name	Family	Conservation status	Occurrence in selected PAs		
					CWS	DDWS	MNP
95	*Phragmites karka* (Retz.) Trin. *ex* Steud.	Dharma, Nalkhagra	Poaceae	LC	✓	✓	
96	*Physalis angulata* L.	Fotka	Solanaceae	LC		✓	
97	*Polygala furcata* Royle		Polygalaceae	V		✓	
98	*Pueraria peduncularis* (Grah. *ex* Benth.) Benth.		Fabaceae	LC		✓	
99	*Raphanus sativus* L.	Mula	Brassicaceae	LC	✓		
100	*Rhynchospora corymbosa* (L.) Britton		Cyperaceae	LC	✓	✓	
101	*Saccharum officinarum* L.	Akh	Poaceae	CD	✓		
102	*Saccharum ravennae* L.	Ekor	Poaceae	DD	✓		
103	*Saccharum spontaneum* L.	Kash, Kaichcha, Kagara	Poaceae	LC	✓	✓	
104	*Sacciolepis indica* (L.) A. Chase		Poaceae	LC		✓	
105	*Sagittaria sagittifolia* L.	Muyamuya, Chhotokut	Alismataceae	LC		✓	
106	*Senna hirsuta* (L.) Irwin & Barneby		Caesalpiniaceae	NE		✓	
107	*Sesbania bispinosa* (Jacq.) Wight.		Fabaceae	LC	✓		
108	*Setaria sphacelata* (Schum.) Stapf. & C.E. Hubb. *ex* M. B. Moss		Poaceae	CD		✓	
109	*Setaria verticillata* (L.) P. Beauv.		Poaceae	LC	✓		
110	*Solanum americanum* Mill.	Tit-begun	Solanaceae	LC	✓	✓	✓
111	*Solanum tuberosum* L.	Golalu	Solanaceae	LC	✓		
112	*Sorghum bicolor* (L.) Moench	Jowar	Poaceae	CD			✓
113	*Thysanolaena maxima* (Roxb.) Kuntze	Jahruful	Poaceae	LC	✓	✓	
114	*Typhonium trilobatum* (L.) Schott	Ghetkul	Araceae	LC	✓		✓
115	*Vernonia cinerea* (L.) Less.	Shial lata, Dankuni, Kuksim	Asteraceae	LC	✓	✓	✓
116	*Xanthium indicum* Koen. *ex* Roxb.	Khagra, Ban-okra	Asteraceae	LC			✓
117	*Xanthosoma sagittifolium* (L.) Schott	Panchamukhi katchu	Araceae	LC			✓

SN	Botanical name	Local name	Family	Conservation status	Occurrence in selected PAs		
					CWS	DDWS	MNP
118	*Xanthosoma violaceum* Scott.	Dudh katchu, Dastur	Araceae	LC			✓
119	*Zea mays* L.	Vuttra	Poaceae	CD	✓		
120	*Zingiber capitatum* Roxb.	Jongli Ada	Zingiberaceae	NE (rare)		✓	
121	*Zingiber officinale* Rosc.	Ada	Zingiberaceae	LC	✓		✓

Table 6.
List of fodder yielding herbs recorded from the selected protected areas [here, DD, data deficient; LC, least concern; NE, not evaluated; NE (rare), not evaluated but seems to be rare; V, vulnerable].

5.4.4 Climbers and other fodder plants

Climbers growing on bushes, trees, and forest floor are important fodder. The leaves and young shoots of the climbers and lianas are mainly eaten by wildlife as food. There were 39 climber plants belonging to 28 genera and 14 families reported as fodder species from the three protected areas. However, review of other reports indicates that the fodder yielding climber composition is almost similar to other forests of southeastern and central regions of Bangladesh. Besides the climber, there were seven fodder yielding pteridophytic species which belong to different genera of seven families. A total of five exotic climbers were recorded to be cultivated by the local people inside the boundary of the protected areas which sometimes raided by wildlife, i.e., monkey, squirrel, etc. The conservation status indicated the presence of one vulnerable plant named *Calamus latifolius*, the fruit of which is eaten by different birds and wildlife as food (**Table 7**).

SN	Botanical name	Local name	Family	Conservation status	Occurrence in the PAs		
					CWS	DDWS	MNP
1	*Acacia pennata* (L.) Willd.	Teorakanta	Mimosaceae	NE (rare)			✓
2	*Ampelocissus barbata* (Wallich) Planch.		Vitaceae	CD			✓
3	*Ampelocissus latifolia* (Roxb.) Planch.		Vitaceae	NE			✓
4	*Basella rubra* L.	Poi shak	Basellaceae	LC			✓
5	*Benincasa hispida* (Thunb.) Cogn.	Chalkumra	Cucurbitaceae	LC	✓		
6	*Caesalpinia digyna* Rottler	Kotchoi Kanta, Umulkuchi	Caesalpiniaceae	LC	✓	✓	
7	*Calamus latifolius* Roxb.	Budum bet, Korak bet	Arecaceae	V	✓	✓	

SN	Botanical name	Local name	Family	Conservation status	Occurrence in the PAs		
					CWS	DDWS	MNP
8	*Calamus tenuis* Roxb.	Chiringbet, Sanchi Bet, Bandari Bet, Jali bet	Arecaceae	LC	✓	✓	✓
9	*Cissus elongata* Roxb.		Vitaceae	LC			✓
10	*Citrullus lanatus* (Thunb.) Matsumura & Nakai	Tormuj	Cucurbitaceae	LC	✓		
11	*Coccinia grandis* (L.) Voigt	Kawa jangi, Telakucha	Cucurbitaceae	LC			✓
12	*Coccinia grandis* (L.) Voigt	Telakucha	Cucurbitaceae	LC	✓	✓	
13	*Cucumis melo* L.	Khira	Cucurbitaceae	LC	✓		
14	*Cucumis sativus* L.	Khira, Futi	Cucurbitaceae	LC	✓		
15	*Cucurbita maxima* Duch. *ex* Lamk.	Mistikumra	Cucurbitaceae	LC	✓		✓
16	*Dalbergia pinnata* (Lour.) Prain	Lalong-chhali, Keti	Fabaceae	LC		✓	
17	*Dioscorea belophylla* (Prain) Voigt *ex* Haines	Dudh alu	Dioscoreaceae	LC			✓
18	*Dioscorea bulbifera* L.	Pagla Alu	Dioscoreaceae	LC	✓	✓	✓
19	*Dioscorea esculenta* (Lour.) Burkill	Maitta Alu	Dioscoreaceae	NE			✓
20	*Dioscorea hamiltonii* Hook. f.	Thakan Budo	Dioscoreaceae	NT		✓	
21	*Dioscorea pentaphylla* L.	Alu lata	Dioscoreaceae	LC	✓	✓	✓
22	*Ipomoea aquatica* Forssk.	Kalmi Shakh	Convolvulaceae	LC	✓		
23	*Ipomoea batatas* (L.) Lamk.	Mistialu	Convolvulaceae	LC	✓		
24	*Lablab purpureus* (L.) Sweet	Shim	Fabaceae	LC	✓		✓
25	*Lagenaria siceraria* (Molina) Standl.	Lau	Cucurbitaceae	LC	✓		✓
26	*Luffa acutangula* (L.) Roxb.	Jhinga	Cucurbitaceae	LC	✓		✓
27	*Luffa cylindrica* (L.) M. Roem.	Purul	Cucurbitaceae	LC	✓		
28	*Mikania cordata* (Burm. f.) Robinson	Assamlata	Asteraceae	LC			✓
29	*Momordica charantia* L. var. *charantia* C. B. Clarke	Karolla	Cucurbitaceae	LC	✓		✓

SN	Botanical name	Local name	Family	Conservation status	Occurrence in the PAs		
					CWS	DDWS	MNP
30	*Momordica cochinchinensis* (Lour.) Sprengel	Kakrol	Cucurbitaceae	LC	✓		✓
31	*Piper betle* L.	Pan	Piperaceae	LC	✓		
32	*Smilax perfoliata* Lour.	Kumari lata	Smilacaceae	LC			✓
33	*Solena amplexicaulis* (Lamk.) Gandhi		Cucurbitaceae	LC	✓		
34	*Tapiria hirsuta* Hook. f.		Anacardiaceae	CD	✓		
35	*Tetrastigma bracteolatum* (Wall.) Planch.	Golgoli lata	Vitaceae	CD			✓
36	*Trichosanthes anguina* L.	Chichinga	Cucurbitaceae	LC	✓		
37	*Trichosanthes dioica* Roxb.	Patal	Cucurbitaceae	LC	✓		
38	*Uvaria hirsuta* Jack	Banor kola	Annonaceae	NE			✓
39	*Vigna unguiculata* (L.) Walp.	Borboti	Fabaceae	LC	✓		
40	*Angiopteris evecta* (Forst.) Hoffm	Dhekia Shak	Angiopteridaceae	LC		✓	
41	*Blechnum orientale* L.		Blechnaceae	LC	✓	✓	
42	*Christella arida* (D. Don) Holtt.	Bish Dhekia	Thelypteridaceae	LC			✓
43	*Diplazium esculentum* (Retz.) Sw.	Dhekia, Dhekia Shak	Athyriaceae	LC	✓	✓	
44	*Helminthostachys zeylanica* (L.) Hook.	Shada Dhekia	Helminthostachyaceae	LC		✓	✓
45	*Lygodium microphyllum* (Cav.) R. Br.		Lygodiaceae	LC	✓		
46	*Marsilea quadrifolia* L.		Marsileaceae	LC	✓		

Table 7.
List of fodder yielding climbers and ferns recorded from the protected areas [here, CD, conservation dependent; LC, least concern; NE, not evaluated; NE (rare), not evaluated but seems to be rare; NT, near threatened; V, vulnerable].

6. Threats to the fodder species

Threats to the fodder yielding plant species are similar to that of the protected areas and forests across the different regions of Bangladesh. According to the fifth report on CBD submitted by the Bangladesh government in 2015, direct threats to the PAs are (1) encroachment in protected areas, (2) degradation of forests and wetlands, (3) infrastructure development, (4) unsustainable and/or illegal exploitation of terrestrial resources, (5) unsustainable and/or illegal fishing practices, (6) change in hydrological regime, (7) pollution, and (8) invasive species [33]. On the other hand, the indirect threats are the institutional and environmental conditions that are behind the direct threats visible on the ground [34]. The main indirect threats are (1) poor institutional capacity, (2) lack of coordination among different agencies, (3) policy and information gaps, (4) lack of enforcement, (5) inadequate and poorly managed system of protected areas, (6) corruption, (7) lack of political commitment, (8) lack of awareness, (9) climate and biophysical changes, and (10) lack of alternative livelihoods in sensitive habitats.

An unprecedented threat to the fodder plant diversity of Bangladesh is exerted by the conversion of wildlife habitat into human settlements along with rampant urban development throughout the country's forested areas [35].

The ecosystem integrity of the PAs of Bangladesh are in very vulnerable situation because they are part of reserved forests which have, in most cases, only been declared after being degraded heavily by various means. It has been assumed that 10% of it is already extinct due to overexploitation. The Bangladesh National Herbarium (BNH) has reported 106 vascular plant species at varying degrees of risk of extinction [33].

The main threats to flora and fauna of protected area conservation emanate from the degradation of forests and wetlands. It is assumed that the anthropogenic pressures on natural resources caused ecosystem depletion. Biotic pressures are exacerbated by dramatic change in climate pattern coupled with sea level rise, increase soil salinity, and increase incidence and severity of cyclones and change in rainfall patterns and temperatures, disturbing the regular seasonality of fruit and flower blooms. This impacts the regeneration of important flora and fauna species and disrupts food chain.

A study revealed that protected areas in the tropical moist evergreen and semi-evergreen forests of hilly regions were highly subject to illegal wood cutting, while those in tropical moist deciduous forests of plain land area were prone to encroachment for settlement and agriculture, and those in mangrove forests of littoral zones were extremely vulnerable to wildlife poaching [36].

7. Effect of fodder plants on wildlife population

Plants and animals are two of the main interacting components of an ecosystem. A very close symbiotic relationship exists between flora and fauna including microbes, i.e., fungi, algae, bacteria, etc. Pollination, decomposition of degradable wastage, nutrient cycling, forming food web, and maintaining the food chain are the main fields where contribution of fauna is very significant. On the other hand, supplying nutrients through food production, providing shade and shelter, and maintaining fertility and productivity of an ecosystem are the key contributions of the floral communities to the associated ecosystem. The smaller animals use the cover of plants and dead leaves to hide from the predators. These animals, i.e., moth, katydids, frogs, and grasshoppers, can blend into the surrounding environment at will and become invisible to the predators searching for food. The tropical rain forests like the protected areas of Bangladesh are very much responsive to animal and plant interaction. In adaptive surroundings of tropical forests, a huge diversity of animals, birds, and insects subsist together. An area of 6 square kilometer of typical tropical rain forest can harbor as much as 1500 flowering plants, 400 species of birds, 100 reptiles, and 60 amphibians along with thousands of butterfly species. However, in a complex ecosystem, the biotic interactions can be toward

DOI: http://dx.doi.org/10.5772/intechopen.85205

any directions, i.e., plant-plant, plant-animal, plant-microbe, animal-microbe, and animal-animal. All of these interactions employ different biotic services [37].

The richness of a wildlife habitat with fodder very sharply determines the fluctuation of wildlife population. Unavailability of food inside the forests make many of the wildlife to come out toward adjacent localities in search of food. For example, a higher number of human-elephant conflicts (HEC) were reported from the southeastern Bangladesh due to degradation and fragmentation of elephant habitat which caused reduction of Asian elephant's population from 500 in the middle of the last century to 228–327 [38]. It is worth and interesting to mention that in some protected areas like Nijhum Dweep National Park, the deer population is shrinking gradually due to higher competition with the thousands of buffalos and cows for fodder. Similar to that of the Asian elephant and deer, the population of monkey along with other herbivores is also reducing at alarming rate due to degradation of overall habitat quality in Bangladesh [39, 40].

8. Protection and enrichment strategies for fodder plants

The process of conserving rare plant species can be divided into three phases: (i) Identification-determining which species are in danger of extinction. (ii) Protection-determining and implementing the short-term measures necessary to halt a species' slide to extinction. (iii) Recovery-determining and implementing the longer-term measures necessary to rebuild the population of the species to the point at which it is no longer in danger of extinction [41].

People living in and around a forest depend on forest resources for a substantial proportion of their subsistence, including food, fiber, medicines, and other uses [42]. Many others perceive forest exploitation as a means of escaping poverty [43, 44]. Forest conservation is likely to be low on these peoples' priorities if it limits their possibilities for livelihood support [45]. So, the development of living condition of the surrounding local people of the important wildlife habitats through improving their livelihood security and diversifying income, so as to meet all their basic needs, should be of first priority; otherwise the conservation effort will go in vain. It is important to extend and strengthen the protective measures by local administrative bodies of Bangladesh Forest Department (FD) against the threats like settlement, agricultural expansion, severe grazing, hunting, shooting, illegal cutting, etc. Local administrative units (beat offices) of the Forest Department must be strengthened with necessary manpower, staff quarters, equipment, logistics, and training, so that they become more capable to conduct the protection and conservation measures because they are the only authority to look after forest and wildlife.

Steps should be taken to halt further expansion of any agricultural/horticultural practices toward the forests. Awareness raising and consciousness of local people regarding the importance of habitat integrity, wildlife, environmental conservation, biodiversity, and endangered ecosystems are mandatory. The government may provide some incentives through money, small loan, training, etc. to help local people in managing sustainable alternative income-generating programs. Relocation and permanent allocation of some barren lands to the forest-dwelling people may reduce their dependency and threats induced by them on the existing forest. Cultivation of crops, i.e., pineapple, banana, paddy, taro, and lemon, should be restricted to some marginal areas of the forests or protected areas. Cattle grazing and browsing pressure in many protected areas is significant, and hence control of grazing animals for reducing the direct effects of disturbance is essential. Special conservation measures both ex situ and in situ methods may be initiated to conserve the threatened and rare native plant species. Enrichment plantation with native wildlife (i.e. rhesus macaque, capped langur, etc.) fruit-bearing plant species should be conducted in the gap spaces of the forests. Activities that were identified by the IUCN and different conservation organizations which contribute to forest and species conservation, i.e., area-based protection, area-based management, species-centered management, education and awareness, improved law and policy, livelihoods and incentives, and capacity building, may be considered for conservation of forest resources in all protected areas. Permanent sample plots of adequate size (0.5–1.0 ha) may be established in representative vegetation types of each wildlife habitat to facilitate long-term ecological and biodiversity assessment which may help monitor the success of restoration and conservation of the fodder yielding plants. Community patrolling should be strengthened to reduce illicit felling as well as raise awareness among local people regarding nature conservation. Digging furrow in suitable locations across the forest may be helpful for controlling fire infestation and litter extraction by local people using small vehicles.

9. Conclusion

Fragmentation and degradation of wildlife habitat quality resulted in reduction of the diversity and population of fodder yielding plants. It is a worldwide trend in which the policy makers and scientists are concerned about. However, Bangladesh having a substantial area declared as protected areas is trying to conserve and restore the wildlife habitat quality. Still now, the protected areas of the country are still harboring a good number of fodder yielding plants from all habit forms. However, many of the fodder species are having very poor density which may reduce further and get extinct if appropriate species-specific multiplication and conservation measures are not taken immediately.

Acknowledgements

The authors acknowledge the contribution and supports of the Bangladesh Forest Department, Arannayk Foundation, and University of Chittagong during the studies.

DOI: http://dx.doi.org/10.5772/intechopen.85205

Author details

Md Akhter Hossain[1*], Rajib Mahmud[2], Nikhil Chakma[2] and Mohammed Kamal Hossain[1]

1 Institute of Forestry and Environmental Sciences Chittagong University, Chittagong, Bangladesh

2 Food and Agriculture Organization of the United Nations, Dhaka, Bangladesh

*Address all correspondence to: akhter.hossain@cu.ac.bd

References

[1] Baumer M. Trees as browse and to support animal production. In: Speedy A, Pugliese P, editors. Legume Trees and Other Fodder Trees as Protein Sources for Livestock. Proceedings of the FAO Expert Consultation; 14-18 October 1991. Kuala Lumpur, Malaysia: Food and Agriculture Organization of the United Nations; 1991. Available from: http://www.fao.org/docrep/003/T0632E/T0632E01.htm#ch1

[2] Boller B, Greene SL. Fodder crops and amenity grasses. In: Boller B, Posselt UK, Veronesi F, editors. Handbook of Plant Breeding, 5. New York: Springer-Verlag New York; 2010. pp. 13-36. DOI: 10.1007/978-1-4419-0760-8_2

[3] Madibela OR, Kemiso D, Kwedibana J. Quality of wild herbaceous legumes and its role in livestock nutrition. In: Revermann R, Krewenka KM, Schmiedel U, Olwoch JM, Helmschrot J, Jürgens N, editors. Climate Change and Adaptive Land Management in Southern Africa—Assessments, Changes, Challenges, and Solutions. Biodiversity & Ecology. Göttingen & Windhoek: 6, Klaus Hess Publishers; 2018. pp. 202-206. DOI: 10.7809/b-e.00325

[4] Leopold A. Game Management. New York: Charles Scribner's Sons; 1933

[5] Hall LS, Krausman PR, Morrison ML. The habitat concept and a plea for standard terminology. Wildlife Society Bulletin. 1997;**25**:173-182

[6] Somkuwar SR, Chaudhary RR, Mahakhode RH. Study of wild fodder yielding plants in Chandrapur District (M.S.), India. The Botanique. 2012;**16**(2):98-103

[7] Ghosh SB. Biodiversity and wild fodder of Gorumara National Park in West Bengal, India. Journal of Environment and Ecology. 2012;**3**(1):18-35. DOI: 10.5296/jee.v3i1.1940

[8] Khan MAR. Wildlife of Bangladesh: Checklist and Guide. Dhaka: Chayabithi; 2015. p. 568

[9] IUCN Bangladesh. Red List of Bangladesh, Vol. 1: Summary. Dhaka: International Union for Conservation of Nature, Bangladesh; 2015. p. 122

[10] Altrell D, Saket M, Lyckebäck L, Piazza M. National Forest and Tree Resources Assessment 2005-2007 Bangladesh. Dhaka: Bangladesh Forest Department and Food and Agriculture Organization of the United Nations; 2007. p. 178

[11] Hussain Z, Acharya G, editors. Mangrove of the Sundarbans, Vol. 2: Bangladesh. Thailand, Bangkok: Dyna Print and IUCN; 1994

[12] IUCN. Guidelines for Protected Area Management Categories. Cambridge: World Conservation Union (IUCN); 1994

[13] Mulongoy KJ, Chape SP, editors. Protected Areas and Biodiversity: An Overview of Key Issues. Montreal, Canada/Cambridge, UK: SCBD/UNEPWCMC; 2004. p. 52

[14] Lewis C, editor. Managing Conflicts in Protected Areas. Gland, Switzerland: Keystone Center and IUCN; 1996. p. 100

[15] Kanowski PJ, Gilmour DA, Margules CR, Potter CS. International Forest Conservation: Protected Areas and beyond. Canberra: Discussion Paper for IFF; Environment Australia. 1999. p. 52

[16] Halladay P, Gilmour DA, editors. Conserving Biodiversity Outside Protected Areas: The Role of Traditional Agro-Ecosystems. Gland, Switzerland: IUCN; 1995

[17] Hale P, Lamb D, editors. Conservation Outside Nature Reserves. Brisbane: Centre for Conservation Biology, University of Queensland; 1997

[18] Scherr SJ, White A, Kaimowitz DA. New Agenda for Forest Conservation and Poverty Reduction-Making Markets Work for Low-Income Producers. Cambridge/Washington DC/Bogor: Forest Trends/CIFOR/IUCN; 2004. p. 35

[19] Chape S, Blyth S, Fish L, Fox P, Spalding M, editors. United Nations List of Protected Areas. Switzerland, Cambridge, UK: IUCN and UNEP-WCMC; 2003. p. 44

[20] Tuxill J, Nabhan GP. People, Plants and Protected Areas: A Guide to in-Situ Management. London: Earthscan; 2001. p. 248

[21] Mukul SA, Rashid AZMM, Khan NA. Forest protected area system and biodiversity conservation in Bangladesh. In: Mukul SA, Rashid AZMM, editors. 2017. Protected Areas: Policies, Management and Future Directions. New York: Nova Science Publishers; 2017. pp. 157-177

[22] Chowdhury MSH, Koike M, Muhammed N. Embracing collaborative protected area management for conservation: An analysis of the development of the forest policy of Bangladesh. International Forestry Review. 2009;**11**:359-374

[23] Muhammed N, Koike M, Sajjaduzzaman M, Sophanarith K. Reckoning social forestry in Bangladesh: Policy and plan versus implementation. Forestry. 2005;**78**:373-383. DOI: 10.1093/forestry/cpi045

[24] Alam M. Evolution of forest policies in Bangladesh: A critical analysis. International Journal of Social Forestry. 2009;**2**(2):149-166

[25] Muhammed N, Koike M, Haque F, Miah MD. Quantitative assessment of people-oriented forestry in Bangladesh: A case study in the Tangail forest division. Journal of Environmental Management. 2008;**88**:83-92

[26] Mukul SA, Quazi SA. Communities in conservation: Protected area management and enhanced conservation in Bangladesh. In: Proceedings of the Internal Conference on the Future of Forests in Asia and the Pacific Outlook for 2020; 16-18 October; Chiang Mai, Thailand: Asia-Pacific Forestry Commission. 2007. pp. 143-159

[27] Bangladesh Forest Department. Management and conservation: Protected Areas-Introduction [Internet]. 2018. Available from: http://bforest.gov.bd/site/page/5430ce33-561e-44f6-9827-ea1ebaa2c00d/Introduction [Accessed: December 15, 2018]

[28] Hossain MK, Hossain MA. Biodiversity of Chunati Wildlife Sanctuary: Flora. Bangladesh: Arannayk Foundation and Bangladesh Forest Department; 2014. p. 176

[29] Feeroz MM, Hasan MK, Hossain MK. Biodiversity of Protected Areas of Bangladesh, Vol. II: Dudhpukuria-Dhopachari Wildlife Sanctuary. Bangladesh: BioTrack and Arannayk Foundation; 2010. p. 214

[30] Hossain MK, Hossain MA, Feeroz MM, Hasan MK. Biodiversity of Madhupur National Park. Dhaka: Bangladesh Forest Department; 2015. p. 134

[31] Ahmed ZU, Begum ZNT, Hassan MA, Khondker M, Kabir SMH, Ahmad M, et al., editors. Encyclopedia of Flora and Fauna of Bangladesh, Vol. 5-12. Dhaka: Asiatic Society of Bangladesh; 2008

[32] Feeroz MM, editor. Biodiversity of Chunati Wildlife Sanctuary: Fauna. Bangladesh: Arannayk Foundation and Bangladesh Forest Department; 2014. p. 199

[33] Department of Environment. The Fifth National Report of Bangladesh to Convention on Biological Diversity [report]; Department of Environment, Ministry of Environment and Forests, Government of the People's Republic of Bangladesh; 2015

[34] USAID. Bangladesh Tropical Forests and Biodiversity Assessment. Washington: Integra Government Services International; 2016. p. 67

[35] Rahman MR. Causes of biodiversity depletion in Bangladesh and their consequences on ecosystem services. American Journal of Environmental Protection. 2015;**4**(5):214-236. DOI: 10.11648/j.ajep.20150405.13

[36] Chowdhury MSH, Izumiyama S. Potential threats, their patterns and extent to the protected areas of Bangladesh: A 'red flag' to biodiversity conservation efforts. In: Chowdhury MSH, editor. Forest Conservation in Protected Areas of Bangladesh. World Forests, Vol. 20. Cham: Springer; 2014. pp. 231-250

[37] Ewel JJ, Hiremath AJ. Plant-plant interactions in tropical forests. In: Burslem DFRP, Pinard MA, Hartley SE, editors. Biotic Interactions in the Tropics: Their Role in the Maintenance of Species Diversity. Cambridge: Cambridge University Press; 2005. pp. 3-34

[38] Wahed MA, Ullah MR, Irfanullah MH. Human-Elephant Conflict Mitigation Measures: Lessons from Bangladesh. Dhaka, Bangladesh: International Union for Conservation of Nature Bangladesh; 2016. p. 30

[39] Ahsan MF, Uddin MM. Human-rhesus monkey conflict at Rampur Village under Monohardi Upazila in Narsingdi District of Bangladesh. Journal of Threatened Taxa. 2014;**6**(6):5905-5908. DOI: 10.11609/JoTT.o3818.5905-8

[40] Green KM. Primates of Bangladesh: A preliminary survey of population and habitat. Biological Conservation. 1978;**13**:141-160

[41] Wilcove DS. Endangered species management: The US experience. In: Sodhi NS, Ehrlich PR, editors. Conservation Biology for all. New York: Oxford University Press; 2010. pp. 220-235

[42] Neumann RP, Hirsch E. Commercialization of Non-timber Forest Products: Review and Analysis of Research. Bogor, Indonesia: Centre for International Forestry Research; 2000

[43] Belcher BM. Forest product markets, forests and poverty reduction. International Forestry Review. 2005;7:82-89

[44] Sunderlin WD, Angelsen A, Belcher B, Burgers P, Nasi R, Santoso L, et al. Livelihoods, forests, and conservation in developing countries: An overview. World Development. 2005;**33**:1383-1402

[45] Ghazoul J, Sheil D. Tropical Rain Forest Ecology, Diversity and Conservation. New York: Oxford University Press; 2010. pp. 371-372

Section 2

Wildlife Diseases

Chapter 6

TSE Monitoring in Wildlife Epidemiology, Transmission, Diagnosis, Genetics and Control

Carla Neves Machado, Leonor Orge, Isabel Pires, Adelina Gama, Alexandra Esteves, Ana Paula Mendonça, Ana Matos, Anabela Alves, Carla Lima, Estela Bastos, Fernanda Seixas, Filipe Silva, João Carlos Silva, Luis Figueira, Madalena Vieira-Pinto, Maria De Lurdes Pinto, Nuno Gonçalves-Anjo, Paula Tavares, Paulo Carvalho, Roberto Sargo and Maria Dos Anjos Pires

Abstract

Among the transmissible spongiform encephalopathies (TSEs), chronic wasting disease (CWD) in cervids is now the rising concern within Europe. CWD will be outlined in this chapter gathering its epidemiology, transmission, diagnosis, genetics, and control. Prion diseases are fatal neurodegenerative diseases characterized by the accumulation of an abnormal isoform of the prion protein (PrP^c), usually designated by PrP^{sc} or prion. CWD is a prion disease of natural transmission affecting cervids detected mainly in North America. The first European case was detected in Norway, in 2016, in a wild reindeer; until April 2018, a total of 23 cases were described. The definite diagnosis is *postmortem*, performed in target areas of the brain and lymph nodes. Samples are first screened using a rapid test and, if positive, confirmed by immunohistochemistry and Western immunoblotting. It is not possible to establish a culling plan based on the genotype, once affected animals appear with all genotypes. However, some polymorphisms seem to result in longer incubation periods or confer a reduced risk. The control is not easy in captive cervids and even more in the wildlife; some recommendations have been proposed in order to understand the danger and impact of CWD on animal and public health.

Keywords: prion, cervids, chronic wasting disease, *PRNP* gene, pathology

1. Introduction

The study and monitoring of wildlife diseases are key points for establishing conservational policies for wild fauna.

Wildlife are often in double risk from disease due to the high number of infectious diseases of wildlife origin that affects humans and (or) domestic animals and also the opposite, humans and domestic animals' diseases that could affect the wildlife. Likewise, due to habitat changes, the introduction of pathogens from domestic animals as well as other actions with long-term adverse effects for the conservation of species needs to have supervising.

As in other diseases, some wildlife animal species could be at risk for transmissible spongiform encephalopathies (TSEs), acting as a potential prion reservoir, threatening the livestock and public health.

TSEs are fatal neurodegenerative diseases characterized by the accumulation of an abnormal isoform, partially resistant to the enzymatic digestion, of the cellular prion protein (PrP^{c}), usually designated by PrP^{sc} or prion. Due to its conformational arrangement, it is very resistant to common inactivation procedures used on conventional infectious agents. As PrP^{c} is host-encoded by the *PRNP* gene, polymorphisms in this gene can act upon the susceptibility or the resistance to TSEs.

The most common and well-known diseases of this group are scrapie in small ruminants, bovine spongiform encephalopathy (BSE), Creutzfeldt-Jakob disease (CJD) in humans, and chronic wasting disease (CWD) in cervids.

Scrapie and CWD are recognized as natural transmitted TSEs, so wildlife can be naturally affected by these two TSEs.

Since scrapie was identified in mouflon sheep (*Ovis musimon*) [1], wild species of sheep and goats, like Iberian wild goat (*Capra pyrenaica*), and Pyrenean chamois (*Rupicapra P. pyrenaica*) can also be infected by scrapie as well as by BSE. Furthermore, according to some studies, European wild ruminants have a *PRNP* genetic background that is compatible with TSE susceptibility [2, 3].

Albeit limited, some countries, namely, Estonia, Finland, Spain, and Cyprus, reported negative results on samples tested for TSE in other wild animal species—mink (*Mustela lutreola*), fox (genus *Vulpes*), raccoon dog (*Nyctereutes procyonoides*), bison (*Bison bison*), and Cyprus mouflon (*Ovis gmelini ophion*) [4].

Considering that CWD is a TSE affecting several cervid species, a very contagious disease with an efficient horizontal transmission, appearing to be enzootic and to be expanding both geographically and in prevalence [5], all the sections in this chapter are focused in order to better characterize the epidemiology, transmission, pathogenesis, diagnosis, genetics, and control of this disease.

2. Molecular basis of transmissible spongiform encephalopathies (TSEs)

2.1 Biology of the etiologic agent

Initially, TSEs were thought to be caused by "slow viruses" (reviewed in [6]). However, as the agent causing scrapie was not deactivated by both chemical and physical procedures, which modify or destroy nucleic acids, it was suggested that this infectious agent was not harboring nucleic acids. Thus, in 1967 Griffith proposed a model in which the scrapie agent could be a protein, but it was Prusiner in 1982, after confirming that procedures used to modify or destroy proteins deactivated the scrapie agent, who published that the etiologic agents of TSEs were proteinaceous infectious particles, called prions [6, 7].

According to this protein-only hypothesis, TSEs are caused by the conversion of the physiological cellular prion protein (PrP^{c}) into a pathogenic misfolded isoform (designated PrP^{sc}) that is able to propagate by recruiting and transforming more PrP^{c}, by an increase in β-sheet structure and a propensity to aggregate into oligomers (reviewed [6, 8]). Moreover, this conformational change confers to PrP^{sc} a greater insolubility in nonionic detergents, high resistance to heat and chemical sterilization, and partial resistance to protease digestion—the truncated

DOI: http://dx.doi.org/10.5772/intechopen.85797

PrP 27–30 kDa (reviewed in [8–10]). Until now, this latter feature has been used for diagnostic purposes, being PrP^{sc} a diagnostic marker for these diseases (see Section 6 Diagnosis of CWD, in this chapter).

PrP^{c} is host-encoded by the *PRNP* gene (see Section 2.3. Prion protein gene, in this chapter) and is normally present in the cell surface in different organs and tissues of mammals and other vertebrates but with high expression levels in the central and peripheral nervous systems. It is a glycosylphosphatidylinositol (GPI)-anchored glycoprotein of 33–35 kDa with a C-terminal globular domain, an N-terminal flexible tail, a single disulfide bond, and an α-helix content. The N-terminal tail includes two charged clusters, the octarepeat region and a hydrophobic domain. At the C-terminus, in the globular domain upstream of the sialylated GPI anchor, there are two N-glycosylation sites [11]. Thus, PrP^{c} has two putative sites of glycosylation, and three glycoforms of PrP can be described: di-, mono-, and non-glycosilated PrP. The relative proportions of these glycoforms and the size of the unglycosylated PrP^{sc} fragment are dependent on the strain of prion (reviewed in [12]).

Regarding the physiological function of the PrP^{c}, it has not been clarified yet; nevertheless, there are several proposed roles; ones are supported by compatible results of different experiments (neuronal excitability; glutamate receptor function, neurite outgrowth; neuroprotection; copper, zinc, iron, and lactate metabolism; and peripheral myelin maintenance), while others are yielding inconsistent results (synaptic transmission and plasticity, memory formation, stabilization of sleep and circadian rhythm, calcium homeostasis, and toxicity elicited by oligomeric species) [11].

While the structure of PrP^{c} is well studied and identified, the structure of PrP^{sc}, the mechanism by which PrP^{c} converts into PrP^{sc} in a posttranslational process, and the molecular mechanisms behind prion strains are still not known, despite all the experimental attempts [8–10].

PrP^{sc}, as a physical template, compels PrP^{c}, with the same primary but different secondary, tertiary, and quaternary structures, to adopt the PrP^{sc} conformation, probably on a complete unfolding of PrP^{c}, followed by refolding. To thoroughly comprehend this molecular process, it is essential to know the architecture of PrP^{sc}. As extensively reviewed by Requena and Wille [9, 10], distinct molecular models have been proposed for PrP^{sc} (27–30 kDa): (1) a four-stranded β-sheet plus two C-terminal α-helices [13, 14]; (2) an antiparallel, intertwined β-helix structure [15]; (3) a parallel β-helix [16]; (4) a fold modeled on the human TATA-box-binding protein containing a five-stranded β-sheet, with the C-terminal α-helices [17]; (5) a "spiral model" with three α-helices of the original structure and four β-strands [18]; (6) a unit consisting of a left-handed four-rung parallel β-helical fold for the N-terminal part and α-helical state for the C-terminal portion, assembled as a trimer [19]; (7) a PrP primary structure based onto the β-helical template and the C-terminal α-helices [20]; (8) a β-helical architecture with a threefold domain in the trimeric unit and α-helical structure for the C-terminal portion [21]; (9) a two-rung β-helix with the C-terminal α-helices [22]; (10) a parallel in-register intermolecular β-sheet (PIRIBS) architecture [23, 24]; and (11) a parallel β-helix extended to the C-terminal portion of the PrP molecule [25].

Nevertheless, none of these models have explained all the experimental results. Recently, methodologies like cryo-electron microscopy and X-ray fiber diffraction have pointed out a four-rung β-solenoid as the basic structural element structure of PrP^{sc} [26], though the available structural information is still limited to determine the PrP^{sc} architecture in atomic details (reviewed in [10]).

To explain strain diversity of prions, the prion-only hypothesis considers that, in the absence of a nucleic acid, the variety of PrP^{sc} conformers and its mixture may result in different prion strains. Each prion strain presents a specific disease phenotype (incubation times, clinical signs, histopathological lesions, PrP^{sc} deposition patterns in the brain, and PrP^{sc} biochemical characteristics) which is exactly preserved upon serial passage within the same host genotype. Nevertheless, the molecular mechanisms by which a range of PrP^{sc} conformers would be produced and selected have not been yet established (reviewed in [8]).

In relation to chronic wasting disease (CWD) prion strains, two prevalent CWD prion strains (CWD1 and CWD2) were identified in North America based on transmission in transgenic cervid (TgCerPrP) mice of several CWD isolates from different cervid species and geographic areas. CWD1 and CWD2 showed distinct incubation time, clinical signs, and neuropathologic profile but with indistinguishable electrophoretic migration patterns of di-, mono-, and non-glycosyl forms of PrP^{sc} [27], reviewed in [28]. These bioassay results showed that elk may be infected with either CWD1 or CWD2 strains, while in deer CWD1/CWD2 strain mixture can be present ([27], reviewed in [28]) (**Figure 1**).

Regarding the CWD prion strain(s) responsible for the outbreaks of the disease in new geographic areas—South Korea, Norway, and Finland—the available data is still limited. The strain characterization of the Korean CWD cases in elk in 2001 and 2004 suggested a single strain responsible for the outbreaks imported from Canada [29], without identifying if it was CWD1, CWD2, or both. The biochemical analysis and immunohistochemistry (IHC) distribution of PrP^{sc} from Norway reindeer revealed a pattern indistinguishable from North America isolates [30]. Remarkably, in CWD-affected moose in Norway, a different phenotype was observed in both PrP^{sc} distribution and biochemical features, suggesting a presence of a different type of CWD prion strain in moose from this country (designated Nor-16CWD) [31] (**Figure 1**).

Novel cervid prion strains have been experimentally generated by adaptation of prions from other species, for instance, scrapie (SSBP1, an American classical scrapie isolate) [28] and BSE [32], demonstrating that cervid species can also be susceptible to other prions. Moreover, CWD prions easily adapt to new species (see Section 4), including sheep, cattle, and squirrel monkeys. Thus, there is a putative risk of development of novel CWD-related prion disease in livestock by grazing in CWD-contaminated pasture (**Figure 1**). Lastly, the ability of the CWD prions to cross the human species barrier has to be further evaluated, but amino acid residues (residues 165–175) in the β2–α2 loop sequence of human PrP^{c} can constitute a species barrier to its conversion by CWD prions [33], reviewed in [28].

2.2 Prions and the deviations in the central dogma of molecular biology

According to the central dogma of molecular biology, first published in 1958 and revisited later [34], heritable information is stored in DNA, expressed as RNA, and translated into protein. Nevertheless, this paradigm has been updated by many aspects of the regulation of gene expression, namely, by the identification and characterization of alternative splicing, alternative promoters, alternative polyadenylation events, and the increasing number of noncoding RNAs (ncRNAs) with critical importance in the regulation of messenger RNA (mRNA) [35] and the discovery of "prions": prion proteins can adopt multiple conformations, at least one of which has the capacity to self-template [36, 37] (**Figure 2**).

2.3 Prion protein gene (PRNP)

The astounding improvement in genetic tools and bioinformatic programs/algorithms and the incredible amount of data deposited freely in the main scientific

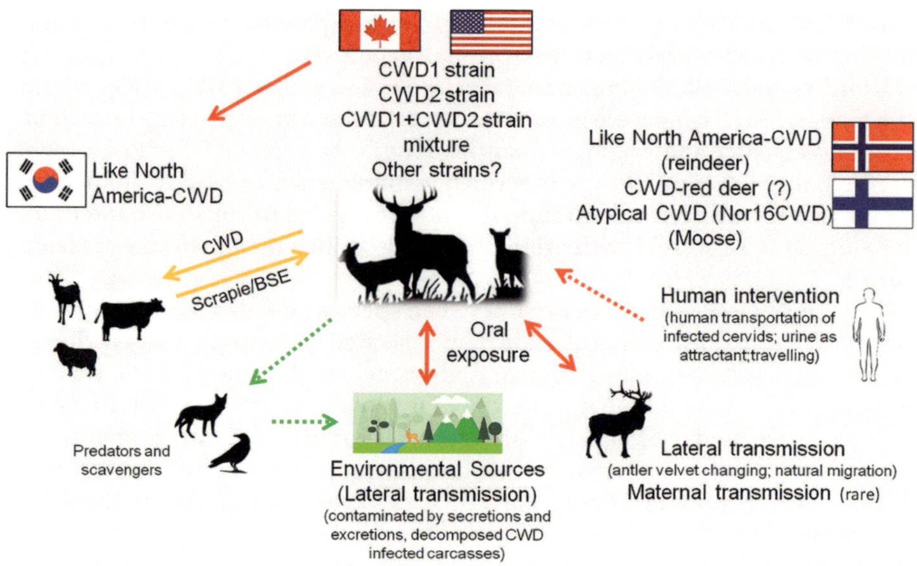

Figure 1.
Summary of CWD prions and transmission. Like North America-CWD-isolates with PrPres biochemical and PrPres distribution similar to that described in North America CWD cases. → natural transmission; ⇢ putative transmission; → experimental transmission; ⇢ potential spread of CWD prions or decrease of PrPres environmental reservoir at the carcass site due to scavenging process, (silhouettes and pictures from freepik.com and img.linkfrog.com).

databases allow us to use comparative genomics in an effective manner. **Figure 3** contains the representation of the *PRNP* gene in *Homo sapiens*, used as a reference and establishing a comparison with other animal species according to the Ensembl database. This gene is constituted by two exons, although a single exon is responsible for the open reading frame (ORF) and the remaining sequence contains untranslated and regulatory regions. Some animal species have the same distribution, while others contain a single exon (e.g., *Canis lupus familiaris*, *Felis catus*, and *Ovis aries*), three exons (*Cervus elaphus* and *Odocoileus virginianus* from *Cervidae* family), or four exons (like *Bos taurus),* according to Ensembl database. Nevertheless, a high level of conservation at the coding sequence and corresponding protein sequence is maintained (as confirmed in **Figure 4**).

The fundamental event in the pathogenesis of TSE is not a primary structure modification but the conversion of the normal cellular prion protein (PrP^{c}) into the misfolded pathogenic isoform (PrP^{sc}) [38]. In fact, PrP^{c} and PrP^{sc} share the same amino acid sequence, but differences in the secondary structure originate the tertiary and quaternary structures that dictate the PrP^{sc} with new physicochemical properties, namely, insolubility in nondenaturing detergents and partial resistance to proteolysis. It is therefore important to identify the specific codons and amino acids with relevant importance in the protein structure dynamics.

Rongyan and collaborators [39] compared the *PRNP* gene sequences among 83 species and reinforced a remarkable degree of conservation among the mammalian sequences. In order to confirm this statement, a DNA, RNA, and protein comparison of *PRNP* among humans, bovine, ovine, caprine, and deer was performed. According to this comparison, human *PRNP* is less similar compared to the others, which is understandable since they are phylogenetically more distant species. However, the protein comparison showed a high similarity between all these species (above 90%), indicating a high conservation of PrP.

In order to add supplementary information especially regarding wild species, the PrP protein sequences from 13 different species were compared and are presented in

Figure 4. The high level of conservation suggests the preservation of some important functional characteristics of PrP through evolution.

According to NCBI-SNP database (accessed in December 2018), 3683 variations in the human *PRNP* gene were presented. Some of these variations are located in the coding sequence and originate modifications in the protein. Once PrP is highly conserved, the variations already described in humans can be used to predict variations in other species. In order to simplify this process, all the missense and nonsense mutations associated with prion diseases described in humans are presented in **Figure 2**.

Variations in PrP sequences exist between species and also between individuals of the same species. It was already demonstrated that this fact can influence the susceptibility to prion infection and ultimately to disease. Chronic wasting disease appear to have a higher potential of transmissibility than other forms of prion disease, and it has been confirmed that some genetic variations are associated with lower rates of infection and slower progression of clinical manifestations [40]. A total of 17 polymorphic sites have been reported in the PrP in *Cervidae* species.

Historically, O'Rourke and collaborators [41] were the first authors that reported that *PRNP* gene from *Cervus elaphus* was polymorphic at codon 132 encoding methionine (M) or leucine (L). This codon is equivalent to codon 129 in

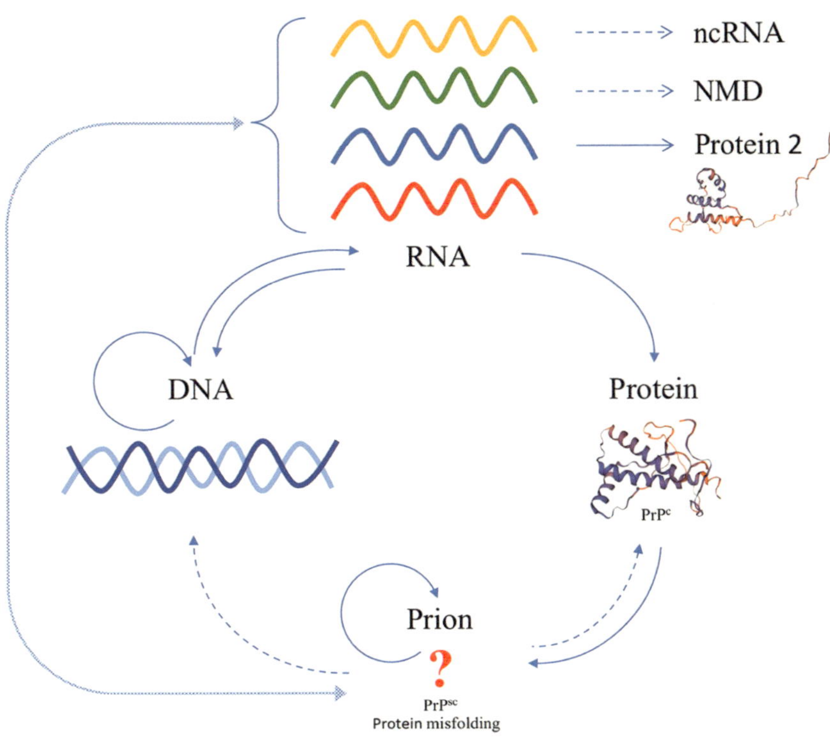

Figure 2.
Updated vision of the classical central dogma of the molecular biology. A DNA sequence can originate multiple RNAs (by using different promoters, by alternative polyadenylation and alternative splicing events). Some of these RNAs can be degraded by nonsense mediated decay (NMD), normally if they contain a premature termination codon (PTC). Other RNAs are not translated but still have a possible regulatory function (noncoding RNAs, ncRNA) and supplementary RNAs are translated originating a different protein, with similar or unrelated function comparing to the canonical protein. The prion postulation assumes that an abnormal protein conformation (PrPsc) propagates itself via an autocatalytic mode by recruiting the normal cellular isoform (PrPc) as a substrate and acting as the disease transmissible agent. This misfolding can be reversible and prion proteins have the ability to interact with nucleic acids (DNA and RNA) and other polyanions (as lipids).

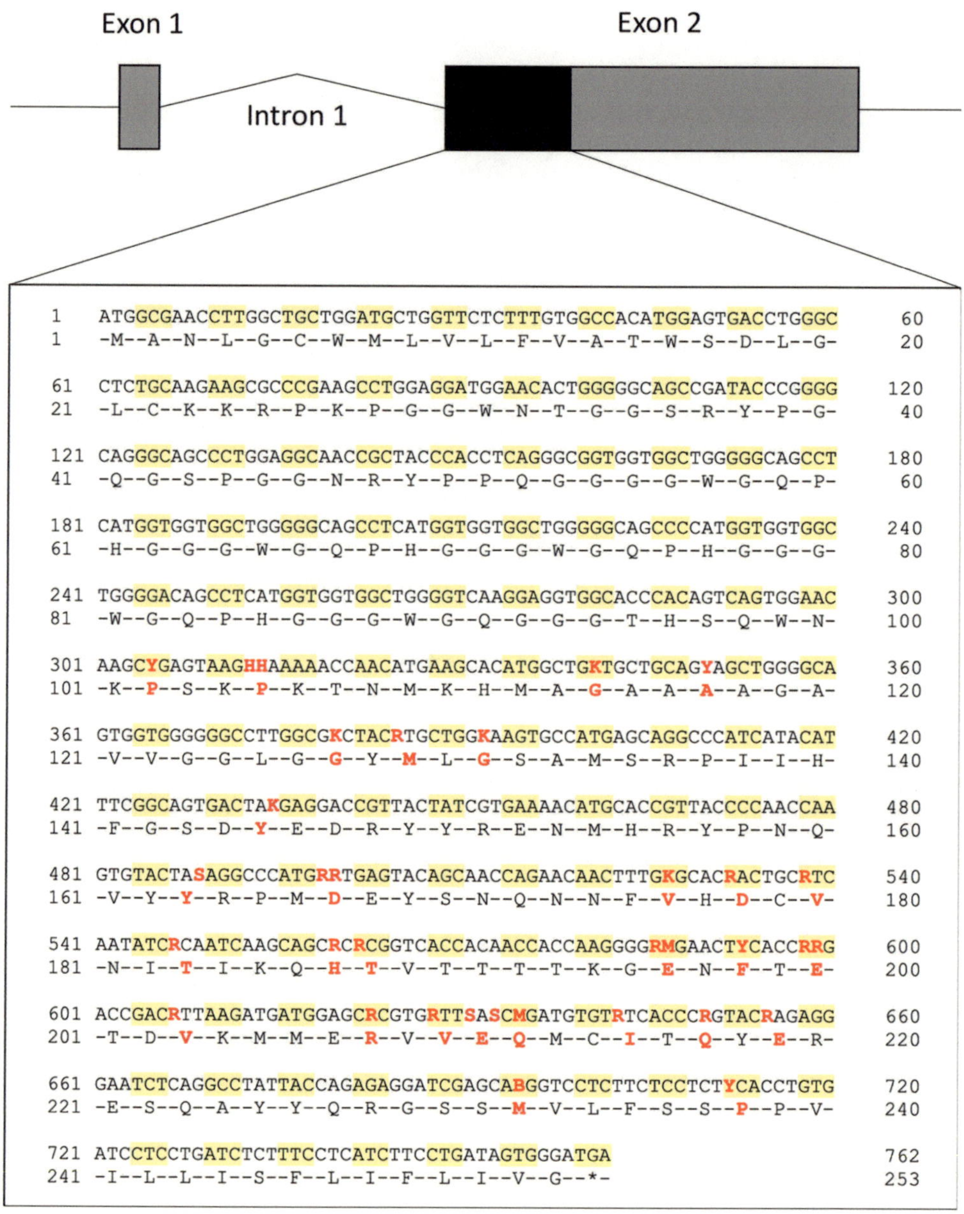

Figure 3.
Organization of the Homo sapiens PRNP gene. From the two exons, only exon 2 is codifying for the PrP protein (black square). The coding sequence is presented below with the top lines showing the nucleotide sequence and the bottom lines showing the amino acid sequence. Missense and nonsense mutations in PrP associated with human prion diseases are marked in red. Information retrieved from Ensembl and NCBI databases were used to create this scheme.

the human protein encoding either M or valine (V) [42]. According to Collinge and colleagues [43] and Palmer and collaborators [44], this variation has an important impact on human prion disease presentation. In the same direction, regarding CWD, O'Rourke and collaborators [45, 46] indicated that the L132 allele protected against this disease in *Cervidae*. Although all *PRNP* genotypes can be affected with CWD, there are some polymorphisms that appear to result in longer incubation periods in some species. The polymorphisms Q95H and G96S are related to the reduction of the risk of infection [47]. S96S or G96S and G95H

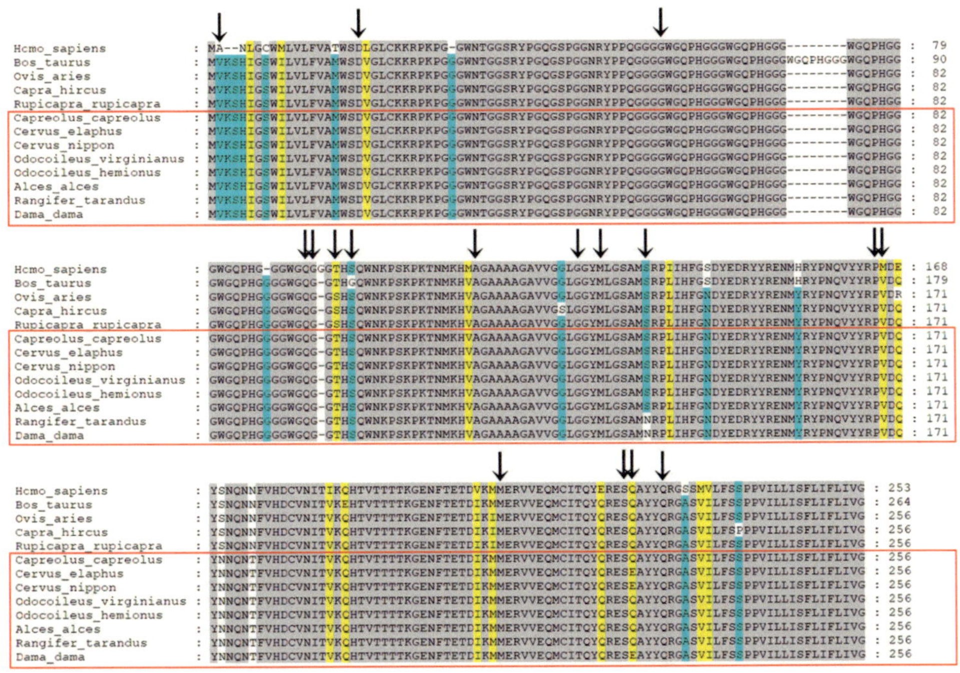

Figure 4.
Alignment of PrP protein sequences among 13 different species. T-coffee was the multiple alignment tool used from EMBL-EBI (https://www.ebi.ac.uk/Tools/msa/tcoffee/) and Genedoc version 2.7.000 was the multiple sequence alignment editor. The accession numbers of each species considering a short name (by order of presentation in the alignment) are: Hs_ENSP00000368752.4; Bt_ENSBTAP00000043233.2; Oa_ENSOARP00000004991.1; Ch_ NP_001301176; Rr_UniProt_spQ5XVM4; Cc_AY639096; Ce_UniProt_ spP67987; Cn_UniProt_trQ6DN38; Ov_UniProt_trQ7JIQ1; Oh_UniProt_trQ6VS46; Aa_UniProt_trQ693S2; Rtg_UniProt_trQ3Y673; Dd_UniProt_trQ7YSF3. Besides human (the reference), the four following species belong to Bovidae family and the last eight species belong to Cervidae family (highlighted with a rectangle frame). The arrows locate the seventeen polymorphic amino acids described in Cervidae.

seem to produce a reduced susceptibility, with longer survivor period [48], being underrepresented in CWD-affected populations [45]. Regarding S225F polymorphism there is a differential susceptibility to experimental oral exposure and incubation periods [45, 49].

Table 1 presents the information concerning the amino acidic variations reported until 2018, with the references and functional implications in the susceptibility to CWD.

3. Epidemiology of chronic wasting disease (CWD)

3.1 Natural history and geographical distribution of CWD

Descriptions of CWD date back to 1967, in which a body-wasting syndrome associated with behavioral changes was described in Colorado in a closed herd of captive mule deer. However, only a few years later, in 1980, the pathologists Elizabeth Williams and Stewart Young observed histological lesions that made it possible to identify the disease as transmissible spongiform encephalopathy, in deer from wildlife research facilities in Colorado and Wyoming. These lesions included neuropil spongiform transformation, intracytoplasmic vacuoles in neuronal perikaryons, and significant astrocytic hyperplasia and hypertrophy [60].

DOI: http://dx.doi.org/10.5772/intechopen.85797

In 1991, amyloid plaques reactive to antibodies produced against scrapie were identified in the cerebral gray and white matter and in molecular, pyramidal, and granular layers of the cerebellum of infected captive mule deer (*Odocoileus hemionus hemionus*) [61]. In the same year, a similar study identified scrapie amyloid-immunoreactive plaques in Rocky Mountain elk (*Cervus elaphus nelsoni*) and hybrids of mule deer and white-tailed deer (WTD) (*Odocoileus virginianus*) [62].

After its first identification, CWD was detected in a mule deer from Wyoming, and, until 1980, they had already identified 53 cases in mule deer (*Odocoileus hemionus hemionus*) and 1 case in a black-tailed deer (*Odocoileus hemionus columbianus*) from wildlife facilities in Colorado and Wyoming [60].

The disease continued to be detected in new cervid species as Rocky Mountain elk (*Cervus elaphus nelsoni*) in Colorado and Wyoming wildlife facilities [63]. Subsequently, the disease was identified in free-ranging animals: mule deer and elk in the same states [64] and white-tailed deer in Nebraska and South Dakota in 2001 [65, 66].

The limited area of southern Wyoming, Northern Colorado, and Western Nebraska has been considered the endemic area. This area spreads rapidly, and new epidemic outbreaks were identified in other states including areas not contiguous geographically to endemic areas [66].

The establishment of disease surveillance and detection programs in both wild and captive cervids contributes to the knowledge of the geographical distribution of the disease [67]. However, the lack of well-defined limits of the affected areas, the low incidence, and the insufficient sampling could lead to underestimated prevalence rates [66].

a.a Position	**Consensus a.a**	**Variation**	**Reference**
2	V	M	[50]
20	D	G	[51, 52]
59	G	S	[3, 52]
95	Q	H	[40, 52, 53]
96	G	S	[52, 53]
98	T	A	[3]
100	S	N	[54]
116	A	G	[52]
129	G	S	[55]
132	M	L	[45, 56, 57]
138	S	N	[55]
168	P	S	[3]
169	V	M	[55]
209	M	I	[50]
225	S	F	[58]
226	Q	E/K	[3, 52, 58, 59]
230	Q	L	[52]

The position considers the Cervidae prion protein. Legend: amino acid codes—A, alanine; D, aspartic acid; E, glutamic acid; F, phenylalanine; G, glycine; H, histidine; I, isoleucine; K, lysine; L, leucine; M, methionine; N, asparagine; P, proline; Q, glutamine; R, arginine; S, serine; T, threonine; V, valine.

Table 1.
Amino acid variations and susceptibility to CWD.

At present, the disease has been identified in captive and free-ranging cervids in 24 states of the United States of America (USA) [68]. Considering only free-ranging cervids, CWD was identified in 270 counties in 24 states, according to March 2019 data, namely, Arkansas, Colorado, Illinois, Iowa, Kansas, Maryland, Michigan, Minnesota, Mississippi, Missouri, Montana, Nebraska, New Mexico, New York, North Dakota, Pennsylvania, South Dakota, Tennessee, Texas, Utah, Virginia, West Virginia, Wisconsin, and Wyoming [69, 70].

In Canada, the first case was detected in 1996 in Saskatchewan, in a captive elk [71]. The second case was diagnosed in 1998 in a farmed elk and the third case, in 2001, in a captive 24-month-old Rocky Mountain elk. In this herd, no other cases had been detected; however, an animal which died 2 years earlier without postmorten examination had been introduced previously from an infected herd [72]. In 2002, the first case was identified in farmed white-tailed deer [71]. Nevertheless, considering a retrospective study done in Toronto Zoo cervids that died between 1973 and 2003, CWD exists in Canada since the 1980s. CWD infection was detected in seven mule deer and one black-tailed deer [73].

According to the Canadian Food Inspection Agency, data about cervid herds infected with CWD are reported since 1996 [69]. In 2001, CWD was identified in 21 herds, the highest number between 1996 and 2018. Until last year, the disease was circumscribed to two provinces: Alberta and Saskatchewan [72]. A screening for CWD in caribou (*Rangifer tarandus)* in northern Quebec revealed no positive animals [74]. However, data from 2018 reported the disease in Quebec in a red deer herd [69, 70, 75].

Recent data, from 2018 official reports, describe CWD in captive elks (in the province of Alberta and Saskatchewan), white-tailed deer (in Saskatchewan), and red deer (in Quebec and Saskatchewan) [69, 75]. CWD was also identified in free-ranging cervid populations in Canadian Alberta and Saskatchewan provinces [69, 71, 75].

The first case outside North America was detected in 2002, in South Korea, associated with a 7-year-old elk imported from Canada [76].

CWD was first recognized in Europe in March 2016 in a free-ranging female reindeer (*Rangifer tarandus*) in the Nordfjella mountain range of Southern Norway [30]. Since then, Norway further intensified its monitoring by sampling with the aid of hunters and governmental culling. A total of 41,125 cervids had been tested leading to the detection of CWD in 19 wild reindeers (*Rangifer tarandus*), 3 wild moose (*Alces alces alces*), and 1 red deer (*Cervus elaphus*) [77, 78]. In Finland, one CWD-affected moose was detected in 2500 cervids tested [78].

3.2 Host range of CWD

CWD has been identified in several members of the family Cervidae: white-tailed deer (*Odocoileus virginianus*) [79], mule deer (*Odocoileus hemionus*) [60], Rocky Mountain elk (*Cervus elaphus nelsoni*) [80], moose (*Alces alces shirasi*) [81], reindeer (*Rangifer tarandus*) [30], moose or Eurasian/European elk (*Alces alces alces*), and red deer (*Cervus elaphus*) [77].

There are other animals that can experimentally be infected, both cervid and noncervid species. The experimental host range includes squirrel monkey [82, 83], several species of voles [84, 85], white-footed mice [29], Syrian golden hamsters [86], ferrets [87], raccoons [88], cats [89], pigs [90], and sheep [91]. Other cervids are susceptible to experimental CWD infection: fallow deer (*Dama dama*) [92] and muntjac deer (*Muntiacus reevesi*) [93]. However, CWD was not yet detected naturally in these species.

DOI: http://dx.doi.org/10.5772/intechopen.85797

3.3 Prevalence and dissemination of CWD

The prevalence of the CWD is quite different in wild or captive cervids. In free-ranging cervids, the disease spreads slowly, often with a stable prevalence and low diffusion rate [94]. However, CWD prevalence in wild cervids and geographical areas with animals infected are increasing every year. Prevalence could reach 31% in hunting deer from Wyoming (USA) [95] and exceed 10% in elk from Rocky Mountain National Park [96]. In Europe, the prevalence in cervids in Norway is 0.05% (23 in 41,125 cervids) and 0.04% in Finland (1 in 2500 cervids), which is very low, but it is expected to increase, as it is a very contagious form of TSE, endemic, and difficult to eradicate [78].

In captive cervids the prevalence is higher and may reach 79% or more of affected animals on the herd. In young animals, this number could be even higher [66, 97]. The annual incidence may be higher than 50% in white-tailed deer and mule deer [95].

The diffusion of CWD, despite the many studies existent, is not well understood. The disease spreads slowly, in free-ranging cervids, probably following river corridors or associated with the movements of males, as proposed for white-tailed deer (*Odocoileus virginianus*) [98]. However, there are no definite conclusions about the origin of the disease in wild cervids. CWD foci in these animals could be extensions of the initial endemic focus, secondary foci, or they could correspond to the spontaneous development of different foci of the disease [94]. In captive animals, CWD spread probably followed slightly different pathways. The occurrence of the disease and its irregular propagation with their appearance in two geographically distant endemic regions must be associated with the commercialization of infected animals between the different herds. Occasionally the disease was detected simultaneously in herds and wild animals in the same region, previously or later, in wild animals suggesting a spillover of the infection from captive to free-ranging animals [94].

4. Transmission

Concerns about transmission between species of TSEs, and particularly transmission of animal TSEs to humans, have existed since the infectious nature of these diseases was demonstrated, but there was no evidence of zoonotic transmission until the link between BSE and a new human TSE (variant CJD or vCJD) was established in the late 1990s. Evidence for the transmission of BSE from cattle to humans came from a diversity of sources, combining careful surveillance and epidemiological studies, often involving the development of new techniques and animal models. The demonstration of the zoonotic transmission of BSE also renewed concerns about the public health risks of other TSEs leading to enhanced surveillance and efforts to control scrapie and CWD [99].

Efficient control measures were applied for BSE and scrapie in cattle, sheep, and goats. However, CWD prions present a challenging risk, affecting both captive and wild cervids [100].

The experimental transmission of BSE in cervids and the bioassay in transgenic mice strongly support the claim that CWD agents are not related to classical BSE. However, the possible relationships between CWD and other TSE agents circulating in animal populations (transmissible mink encephalopathy (TME), scrapie) are still uncertain [99].

TSEs are horizontally transmitted by oral ingestion (kuru, BSE, vCJD, TME, feline spongiform encephalopathy (FSE), CWD), by environmental contact with infectious prions (scrapie, CWD), by blood transfusion (vCJD, CWD, scrapie), or by iatrogenic exposure (vCJD, sCJD) [93]. Horizontal transmission of CWD also includes skin, oral, and nasal routes [101].

CWD prions can be discarded in urine, feces, saliva, blood, and velvet antler (clinical and preclinical stage of the infection) and can be found in skeletal muscle, fat, viscera, and the central nervous system [102, 103]. For this, there is a high risk of exposure to CWD prions, not only for cervids and other wild animals but also for hunters who may contact directly with infected animals and people consuming venison [104] (**Figure 1**).

Although still unclear, natural CWD transmission between animals has occurred. Environmental contamination and infectivity can persist in pastures for 2 years after the death of animals affected with CWD, as shown by studies performed in grazing presumably highly contaminated [49].

Both horizontal and environmental transmissions efficiently disseminate CWD. Decomposing carcasses and the placenta also contribute for this transmission as the prion agent is environmentally very stable persisting for many years where TSE-infected animals stood [94]. Other putative sources for CWD spread are predators/raptors and scavengers as prions remained infectious after passage through the digestive system of American crows [105, 106] and coyotes [107] which were orally infected with scrapie and CWD, respectively (**Figure 1**).

High infectivity and rapid transmission between cervids result in a high prevalence and can exceed 90% in captive animals [108] leading to the increased CWD exposure of humans and other animals, mainly by the consumption of prion-infected animal products or grazing on prion-contaminated pastures, respectively. So, studies concerning the zoonotic risk of CWD and transmission to other animal species are very important with a high interest for public and animal health. The effect of the species barrier (lower transmission efficiency compared to that within the donor species) is very present while studying the interspecies transmission of CWD prions. Investigations identified a key influence of amino acid sequence differences in the prion protein between species as well as a strong effect of the prion strain type [109, 110]. For prions, strain type is characterized by a set of properties that define infection and pathogenesis not explained by differences in amino acid sequence [111]. Both factors influence species interaction, dose, route of inoculation, and age of the host [109, 110]. Furthermore, prion agents can adapt to a new host species, only becoming virulent after repeated passage [112].

For the study of CWD interspecies transmission, experimental inoculations were conducted mainly by intracerebral (IC) inoculation, resulting in a successful transmission to ferrets (*Mustela putorius furo*) [113] but unsuccessful to Syrian golden hamsters (*Mesocricetus auratus*) even after multiple attempts [64]. The ferret-adapted CWD was, however, and after several passages, transmitted to hamsters, demonstrating that interspecies transmission can be altered by the adaptation to a new host [113]. The IC inoculation of mule deer CWD to goats presented an incubation period of about 6 years, a longer period when compared to what happens with scrapie [64, 114].

Transmission experiments of CWD prions from white-tailed deer, mule deer, and elk to transgenic mice expressing deer, elk, sheep, cattle, or human PrP^c suggest that the transmission barrier for CWD prions among different species of the cervid family is low, whereas the transmission barrier for CWD prions to sheep, cattle, and humans is high ([115–119], cited by [114]). Although still unclear, species barrier is thought to be controlled, at least in part, by the differences in the primary PrP sequences of donor and recipient animals [120–122] that affect the structure and folding of each prion protein ([123] cited by [114]). Studies of amino acid differences have demonstrated that there is a high level of conservation of the prion protein between species (90%) being the energy barrier increased for conversion, when compared to others, with some major differences in the amino acid sequence [33].

Although the hypothesis that CWD is transmitted to humans is not supported (reviewed by [100]), results from in vitro experiments suggest that the species barrier is not absolute.

No correlation between CWD prion exposure and human prion disease was demonstrated by epidemiological studies, and humanized transgenic mice did not demonstrate transmission. CWD transmission to macaques has been unsuccessful to date, although squirrel monkeys seem to be susceptible to CWD prions [100].

The negative transmission results reported in seven studies support the conclusion that the transmission barrier associated with the interaction of human PrP and CWD prions is fairly strong and is stronger than the species barrier between human PrP and the BSE prion ([124] cited by [100]).

It is hypothesized that, as a result of an extensive human exposure, a CWD prion strain able to be transmitted to humans can emerge, even with no current evidence that it can cause disease [125]. So there is a risk of exposure for those who directly contact with cervids or contaminated environment or who consume venison [104]. However, variation in CWD prion strains, human genetic heterogeneity, other factors contributing to individual variation, and some evidence that intraspecies and interspecies passage of CWD prions may increase the risk that CWD prions adapt to have a lower species barrier for transmission to humans, complicating this area of research [112].

Wild-type inbred strains of mice (e.g., RIII, C57BL, and VM) were important for the identification and characterization of different types of scrapie prions [126–131], but CWD transmission to wild-type mice presented only few results [29]. In order to optimize time and resources as well as the results obtained, when performing studies of scrapie prions, transgenic (Tg) mice for ovine or cervid prion protein gene were developed [115, 117, 118, 132–134].

According to several studies, cervidized mice have been established as appropriate mouse bioassay models for the study of CWD. CWD prions were transmitted efficiently to TgElk mice with no evidence of transmission to ovinized mice—Tg338 (transgenic mouse line expressing the ovine VRQ prion protein) [119].

There are currently no known cases of interspecies transmission of CWD prions to sheep or goats through natural exposure, and transmission resulting from direct inoculation has been relatively inefficient. Nevertheless, the potential for natural exposure of sheep and goats to CWD prions is presumably rising in parallel with the increasing incidence and geographic range of CWD [119]. In this study transmission of classical scrapie and CWD prions from their native hosts to transgenic mouse lines expressing the ovine (Tg338) or cervid PrP was assessed. Inoculation of transgenic mice with scrapie prions from small ruminants or CWD prions from WTD (white-tailed deer) resulted in distinct transmission patterns. Scrapie prions transmitted efficiently to Tg338 but not to TgElk, whereas CWD prions from a natural host efficiently transmitted to TgElk but not Tg338. Unlike these distinct transmission patterns, efficient transmission of CWD prions following primary passage in sheep was observed in both Tg338 and TgElk mice. Experimentally transmitted CWD in small ruminants has been difficult to distinguish from natural cases of classical scrapie [119].

Peripheral lymphoid tissues may be a target for accumulation in cross-species transmission as observed with Tg338 mice intracerebrally inoculated with elk CWD prions with high rates in the spleen but low doses in the brain [110]. Transmission of CWD from mule deer or elk to sheep has been achieved following intracerebral inoculation [91, 135]. Yet, inoculation with brain homogenate from WTD with

CWD did not result in brain accumulation of PrP^{sc} during primary passage in ovinized mice (Tg338). One factor that may have contributed to the lack of transmission could be varying levels of PrP^{sc} and PrP^{CWD} between brain homogenates from animals with scrapie and CWD, respectively [119].

Non-Tg mice have been described as resistant to CWD infection [115], while the VM/Dk inbred strain of mice infected with wapiti CWD prions reported limited infection [29], suggesting the existence of some variable species barriers to the transmission of CWD [77].

Nalls et al. [93] have demonstrated mother-to-offspring transmission and disease progression to a viable cohort of offspring that were born to CWD-infected dams.

Several researchers noted that dissimilarity between PrP sequences alone does not accurately predict the transmissibility between species and other factors are under investigation. One that was used in recent studies was the conformational selection model, which proposes that compatibility between the three-dimensional molecular shapes of PrP^{sc} and PrP^{c} is more important than the sequences of the protein backbones [136].

5. Pathogenesis of CWD: the interaction of prions and immune system

The unique features of prions as infectious agents are also blatant when considering its interaction with the host's immune system. As stated previously, PrP^{c} is involved in important neurologic functions; however, the depletion of neuronal PrP^{c} in experimental models, such as mice, cows, and goats, did not result in altered phenotypes [11]. This suggests the existence of redundant mechanisms that compensate the absence of PrP^{c}, albeit its ubiquitous cellular expression. Pertaining the immune system, PrP^{c} is expressed in lymphocytes, granulocytes, and stromal cells. This abundant expression may explain why PrP^{sc} does not elicit specific immune responses despite its accumulation in high levels within the host's secondary lymphoid tissues (SLT) in the initial phases of the infection. In particular, gut-associated lymphoid tissues (GALT), such as the Peyer's patches in the small intestine, seem to be pivotal for the infection and disease process. As in most natural occurring prion diseases, CWD of cervids is acquired peripherally, mainly after oral exposure, after which prions replicate in GALT. The importance of GALT in the pathogenesis of prion diseases is well documented by several studies that revealed impaired neuroinvasion in the absence of Peyer's patches [137–140].

5.1 *M* cells

Within GALT, studies suggest that prions depend on host *M* cells to enter the organism, as they seem to be involved in the initial transfer of the prion agent across the gut epithelium. These are specialized epithelial cells widely spread among the follicle-associated epithelium, which are involved in the transcytosis of particulate antigens and microorganisms from the gut lumen into the GALT [141]. *M* cells are essential to the gut immune system being fundamental to antigen recognition and presentation of pathogens or commensal microflora. As such, factors that increase *M*-cell density in the gut epithelium (coinfections or miscellaneous inflammatory conditions) may also increase susceptibility to orally acquired prion infections. A study by Wyckoff et al. [142] mentions the binding of prions to certain types of soil particles that may result in an increased uptake of prions from the gut lumen, but it is not yet clear if this uptake is mediated by *M* cells. Host infection by

M-cell-independent mechanisms cannot be disregarded, namely, those involving directly enterocytes from the follicle-associated epithelium [143] or other *M*-cell-independent pathways [144].

5.2 Follicular dendritic cells (FDC) and B and T lymphocytes

Several studies show that in prion-infected hosts, high accumulations of disease-specific PrP are detectable within the B-cell follicles of the SLT [145–149]. After internalization, evidence suggests that prions accumulate and replicate in the follicular dendritic cells (FDC) that lie in the B-cell follicles and germinal centers of SLT [150–156]. FDC possess multiple slender and long dendritic processes, differentiate from ubiquitous perivascular precursor cells (pericytes), and are a distinct lineage from bone marrow-derived conventional dendritic cells (DC) [157, 158]. FDC extend throughout the B-cell follicle being able to trap and retain large amounts of native antigens upon their surfaces in the form of immune complexes composed by antigen–antibody and/or opsonizing complement components. Apparently, prions can be bound by opsonizing complement components [152–155], internalized, and replicated within FDC, as it was demonstrated that they accumulate upon PrP^{c}-expressing FDC in the SLT of experimentally infected mice, sheep with natural scrapie, cervids with CWD, and patients with vCJD [145–149]. Subsequently, immunohistochemistry studies identified FDC as the first sites of prion conversion and replication in experimentally infected mice, in which increased prion labeling was observed on the plasma membranes of FDC in the Peyer's patch germinal centers [143]. Furthermore, evidence suggests that FDC maturation and regression cycle are impaired by disease-associated prion protein, thus affecting immune function [159]. Despite this, some prion strains can establish infection within SLT regardless of FDC, as it was demonstrated by experimental studies in mice, where infection occurred associated to the high endothelial venules in the lymph nodes [160].

The role of B and T cells in prion disease pathogenesis remains uncertain. Experimental studies showed that mice lacking mature B and T cells were not susceptible to peripheral infection with prions [145, 161–163], despite prion disease pathogenesis being unaffected in T-deficient mice [162, 164, 165]; however, it seems that prions are not capable of replicating within B and T cells [166, 167]. These cells appear to be intermediary actors in prion diseases, by providing physical and humoral support to other cells, thus contributing to the homeostatic state of the immune system. However, B cells, in particular, seem to play a more active role during prion infection. The ability of secreting tumor necrosis factor-α (TNF-α) and lymphotoxins (LT) turns B cells pivotal in keeping FDC in their differentiated state, since in the absence of stimulation from these cytokines, FDC rapidly dedifferentiate [168]. In Peyer's patches, B cells migrate to the mesenteric lymph nodes and then return to the circulation [169]. It is this constant rotation between Peyer's patches and SLT that may help B cells to disseminate prions acquired from FDC as they both migrate together through the germinal centers [170, 171]. Additionally, the accumulation of prions in the spleen and subsequent neuroinvasion are significantly decreased in the specific absence of B cells [162]. It is also noteworthy that prions were detected in B cells from the blood of sheep with scrapie [172] and deer with CWD [173].

As it was already mentioned, FDC in the Peyer's patches are major sites of early prion replication in orally infected animals, both experimentally [137–140, 174] and in the natural forms of the disease. The study of natural scrapie cases in sheep showed that GALT of the upper gastrointestinal tract exhibited early accumulation of prions when compared to the large intestinal GALT, such as the cecal patches [175–178]. The rectoanal mucosa-associated lymphoid tissues (RAMALT), which also accumulate prion in naturally occurring cases of scrapie in sheep and goats, as well as in CWD in cervids, also displayed reduced incidence of prion accumulation at the earlier stages of disease [96, 179–184]. Taken together, these studies point out to the fact that prions first replicate within the GALT of the upper gastrointestinal tract and then spread to the local lymph nodes and to other SLT, such as the spleen. Additionally, it was observed that FDC in the spleen expressing PrP^{c} were able to display high levels of prion replication that was impeded after PrP^{c} expression blockage in these cells [185].

5.3 Mononuclear phagocytic cells

A heterogeneous population comprised of monocytes, conventional dendritic cells (DC), and macrophages seem to have also an important role in host infection. From a different lineage of the stromal-derived FDC, conventional DC [157, 158, 186], the antigen-presenting cells par excellence, are also involved in the transport of antigens both within Peyer's patches and toward the mesenteric lymph nodes [187–189]. Their strategic position allows them not only to sample their local environment for pathogens and their antigens, processing them, but also to capture and retain unprocessed (native) antigens [190, 191]. Processed or native antigens are then delivered to B and T cells as DC undergo maturation and migrate toward the local SLT to initiate an immune response [190–193]. Prions are wrapped by DC either by complement opsonization, namely, by C1q and C3 components [155, 194], or in a non-specifically manner, through fluid-phase micropinocytosis [144]. Specific chemokines have been reported to play important roles in prion infections by regulating FDC and DC migration within the SLT. CXCL13 chemokine is expressed by FDC and other stromal cells in the B-cell follicles of SLT and recruits CXCR5-expressing cells toward them [195, 196]. On the other hand, CXCL13-CXCR5 signaling mediates the migration of certain populations of DC toward the FDC-containing B-cell follicles [192, 193, 197], and studies have shown that the early accumulation of prions upon FDC in Peyer's patches was impaired and disease susceptibility reduced, in the absence of CXCR5 expression by DC [198]. The interplay between FDC and DC during prion infection was also patent by the finding that early replication of prions upon FDC in the local SLT was inhibited when conventional DC were transiently depleted at the time of exposure [199–202]. Apparently, after being transferred across the gut epithelium either directly by enterocytes or by *M* cells, prions may be then internalized by DC [143, 199] and propagated by them toward FDC in Peyer's patches [198]. The ability of DC to migrate into B-cell follicles [192, 193, 203] suggests that they may also propagate prions to and within SLT, namely, to the mesenteric lymph nodes.

Regardless of FDC and DC roles in the propagation of prions, there are other cell populations that appear to phagocytose and destroy them [204, 205]. In fact, it was observed that tingible body macrophages of the germinal centers display heavy PrP^{sc} accumulations within their endosomal compartments during prion disease [143, 159, 185]. It is speculated that macrophages may scavenge and degrade prions in an attempt to protect the host from infection, as it is suggested by studies that show that macrophage depletion resulted in an enhanced accumulation of PrP^{sc} within SLT [206, 207].

5.4 Neuroinvasion

After being acquired orally, prions undergo replication and accumulation upon FDC, reaching a threshold above which neuroinvasion occurs. Experiments in which immunohistochemical tracing was performed revealed that prions then infect enteric nerves and spread along efferent fibers of both the sympathetic (e.g., splanchnic nerve) and parasympathetic (e.g., vagus nerve) nervous systems, spreading within them to the CNS, where they ultimately cause neurodegeneration [143, 208–212]. The initial transfer of prions from FDC to the peripheral nerves continues surrounded by knowledge gaps, but it appears that tunneling nanotubes (TNT) may play an important role at this stage of the disease. These are slender membrane-bound cylinders of cytoplasm that connect cells to enable cell-to-cell communication and the intercellular transfer of plasma membrane or cytoplasmic components [213]. Since mononuclear phagocytes and DC are in close proximity with enteric nerves [214–217] and in vitro studies showed that prions can be transferred between DC and neurons via endolysosomal vesicles within TNT [215–218], it is possible that the same transfer also occurs in vivo. Another hypothesis is that the transfer of prions may also occur in association with small endosomal-derived vesicles, termed exosomes [219], as it was demonstrated experimentally. Studies supporting this theory are based on the observation that DC infected with prions can release exosomes that in turn may infect neighboring cells [220]. In the CNS, prion aggregates accumulate to a point where they overcome the structures within it, starting from the synapses and then proceeding to axons, where axonal retraction is observed, and then to the cell body, in which atrophy and neuronal loss are lastly observed [221–224]. The abovementioned process is, however, slowly and progressive, depending on factors such as prion agent strains, the existence of concurrent infections, and hosts' age and immune status [144].

The onset of neurodegeneration is preceded by the activation of glial cells, microglia, and astrocytes [225, 226]. Microglial cells, in particular, by intervening in synaptic remodeling and in the removal of dead and dying cells, are important for the course of neurological changes during prion infection. Organotypic cerebellar culture devoid of microglia exhibited enhanced prion accumulation [227], whereas CNS prion disease pathogenesis was exacerbated in vivo in the absence of microglia [228]. Studies revealed that during prion infections, a cytokine response mediated by TGF-β and PGE2 develops within the CNS, stimulating microglia to adopt an anti-inflammatory status [223, 229, 230]. Evidence suggests that this anti-inflammatory ambiance may be beneficial for the host, as CNS prion infection in mice deficient in the anti-inflammatory cytokines interleukin (IL)-4, IL-10, and IL-13 displays an aggravated disease pathogenesis [231, 232]. Other studies supported this hypothesis by the finding that blocking colony-stimulating factor (CSF) receptor-1 resulted in the expression of selective anti-inflammatory microglial markers, slowing the development of neuropathology and extending survival times [233]. Activated microglia scavenge and clear prions and prion-affected cells, but neural degeneration seems to occur when prion accumulation is associated to an inflammatory response in the microglia [234]. Together, these data point out to the existence of both anti-inflammatory and pro-inflammatory responses by microglial cells during CNS prion infections [226, 234] contributing to the symptoms of this disease.

6. Diagnosis of CWD

Classification of CWD as TSE required histopathological examination of brains from diseased animals [60]. These subsequent analyses [80] were limited to microscopic evaluation of the CNS to detect neuropathological features typical of, but not necessarily exclusive to, prion diseases including neuronal vacuolation, attendant spongiform degeneration of the neuropil, reactive astrocytic gliosis, and florid amyloid plaques.

6.1 Clinical signs and macroscopic lesions

The main clinical features of progressive CWD disease in adults is weight loss and behavioral changes that typically span weeks or months. Besides weight loss, diseased animals may show a wide range of behavior changes for a long time that include somnolence, lassitude, repetitive walking, polydipsia, and polyuria. Gradually, the affected cervids may exhibit ataxia, mainly in the hind limbs, head tremors, drooped ears, and fixed gaze, and may have difficulty in swallowing. In the end stage, they may show signs of sialorrhea and teeth grinding [235, 236]. The clinical signs may be subtle requiring that the animal keeper/observer is familiar with and has some knowledge of the normal behavior of the species. As a result, the disease is most commonly detected in emaciated cervids hunted, killed, or injured by road accidents or found dead [235].

At necropsy, severe emaciation, poor hair coat condition, megaesophagus, froth or watery rumen contents (often containing sand), abomasal or omasal ulcers, serous atrophy of bone marrow and pericardial fat, enlarged adrenal glands, muscle atrophy [237], and aspiration pneumonia are common findings. As the clinical signs and the macroscopic findings at necropsy are not pathognomonic [49], other diseases must be ruled out, namely, listeriosis, meningoencephalitis, brain abscesses, starvation, nutritional deficiencies or severe parasitism, bluetongue disease, epizootic hemorrhagic disease, meningeal worms (*Parelaphostrongylus tenuis*), and locoweed intoxication [49, 235, 238].

TSEs present very specific diagnostic challenges because of their strain variation, their very long incubation period, and the lack of pathognomonic clinical signs. Moreover, the disease pathogenesis can range both among and within species due to the influence of strain, host genotype, or a combination of the two [239].

6.2 Rapid screening testing

The initial diagnosis of TSE in all species was based on passive surveillance (e.g., clinical presentation) and assessment via conventional histopathology and subsequently detection of PrP^{sc} by IHC, both time-consuming, technically demanding, and expensive for large surveillance plans [77]. For that reason, the development of commercial rapid immunologically based screening tests (RT) for the detection of PrP^{sc} in tissue homogenates allowed a large-scale active surveillance program for TSE in cattle and small ruminants.

These were originally developed for the detection of BSE in the cattle brain, and extensive formal test evaluation of a wide range of these RT was undertaken at Europe level [240, 241]. Mainly, there are two types of these RT kits allowing the purification, concentration, and detection of PrP^{sc} from samples of tissues obtained from infected animals. Both assays consisted in an immuno-enzymatic technique: one is a sandwich format using two antibodies for the detection of the antigen PrP^{sc} after proteinase K digestion (e.g., TeSeE BioRad®), and the other uses a PrP^{sc}-specific ligand immobilized on the surface of the antigen-capture plate (e.g., HerdCheck IDEXX®). Both are suitable for the surveillance of CWD as diagnostic methods for the detection of the disease [77].

The limited experience in CWD so far in Europe has not allowed us to know the best tissue to be proposed as the most sensitive for surveillance in all circumstances. Sampling only the brain stem could reduce the diagnostic sensitivity for strain-host combinations that are characterized by early lymphoid PrP^{sc} accumulation. Consequently, lymphoid tissues (preferably the tonsils, retropharyngeal lymph nodes, and lymphoid tissues of rectal mucosa) and the brain stem at the level of the obex should be tested to maximize the diagnostic sensitivity in any surveillance program [77].

6.3 Confirmatory testing

Like in other recognized animal TSEs, confirmation of an initial CWD "screening positive" (or "suspect") sample can be undertaken by histopathological examination and immunodetection methods either by IHC or Western blot (WB). The former enables PrP^{sc} accumulations to be assessed regarding the types of deposition and anatomical distribution (including cellular) location, while the latter gives some classification data based on the molecular mass and glycosylation profiles of the PrP^{sc} [77].

Histopathologic lesions in the encephalon are similar to those described for ruminant TSEs: perikaryon neuronal vacuoles, microcavitation of the gray matter, astrogliosis, neuronal degeneration and loss, and PrP positively labeled prion deposits and plaques [242]. Duration of clinical disease does not significantly affect the distribution or severity of lesions, and inflammatory cell response is not apparent, unless associated with intercurrent disease [49]. In clinically affected cervids, examination of well-fixed medulla oblongata at the level of the obex is considered sufficient for diagnosis of CWD [80], and sections at this level were used for CWD surveillance before availability of immunohistochemistry [49].

Neuropathology varies slightly between deer and elk: elk have more severe lesions in the thalamus and in some white matter areas. Congo red birefringent and PAS-positive amyloid plaques have been seen in the deer brain but not in elk [80]. The cerebral cortex and basal ganglia of the elk with CWD show minimal fine spongiform degeneration and astrogliosis with focal distribution. The spongiform degeneration with astrogliosis is more prominent in the thalamus where it forms clusters of coarse vacuoles. Fine spongiosis, often in small clusters, is present in the molecular layer of the cerebellum, in dorsal nuclei of the pons, and in the substantia gelatinosa of the spinal cord. Occasional large neurons in various nuclei of the pons show a vacuole [243]. Amyloid plaques are relatively common and can be detected on hematoxylin and eosin (HE)-stained brain sections, most prominently and with decreasing frequency, in white-tailed deer, mule deer, and elk [49]. Neuronal loss and astrogliosis are minimal except for the molecular layer of the cerebellum, which shows rarefaction of granule cells with no indication of apoptosis [243].

The PrP^{sc} immunostaining is consistently present in the cerebral cortex, basal ganglia, and thalamus. In the cerebellum the immunostaining is present in both molecular and granule cell layers as well as in the dentate nucleus. In the pons it is widespread over gray structures, whereas in the spinal cord, it is generally confined to the dorsal part of the dorsal horns [243]. Patterns of PrP^{sc} deposition in CWD-affected cervid brains include perineuronal and perivascular accumulation, extracellular plaques and granular deposits, and subependymal and subpial deposition [49] (**Figure 5**).

Recently, a different neuropathologic phenotype, characterized mainly by intraneuronal deposition of PrP^{sc} and few immunostaining at the dorsal motor of the vagus nerve, was observed in the putative atypical CWD detected in Norwegian moose [31].

Deposition of PrP^{sc} occurs widely in lymphoid tissues (**Figure 5**) during CWD incubation in the absence of histologic lesions in these tissues, resembling classical scrapie in small ruminants [149, 244]. Therefore, lymphoid tissues are very useful for diagnostic purposes and surveillance. Nevertheless, in the referred atypical CWD, no PrP^{sc} was detected in lymphoid tissue [31], similar to that described in heterozygote ARR sheep affected with classical scrapie as well as atypical scrapie. In those cases, lymphoid tissues infectivity should be further studied like it was in atypical scrapie, demonstrating that infectivity can accumulate in lymphoid tissues even with no detectable PrP^{sc} [245].

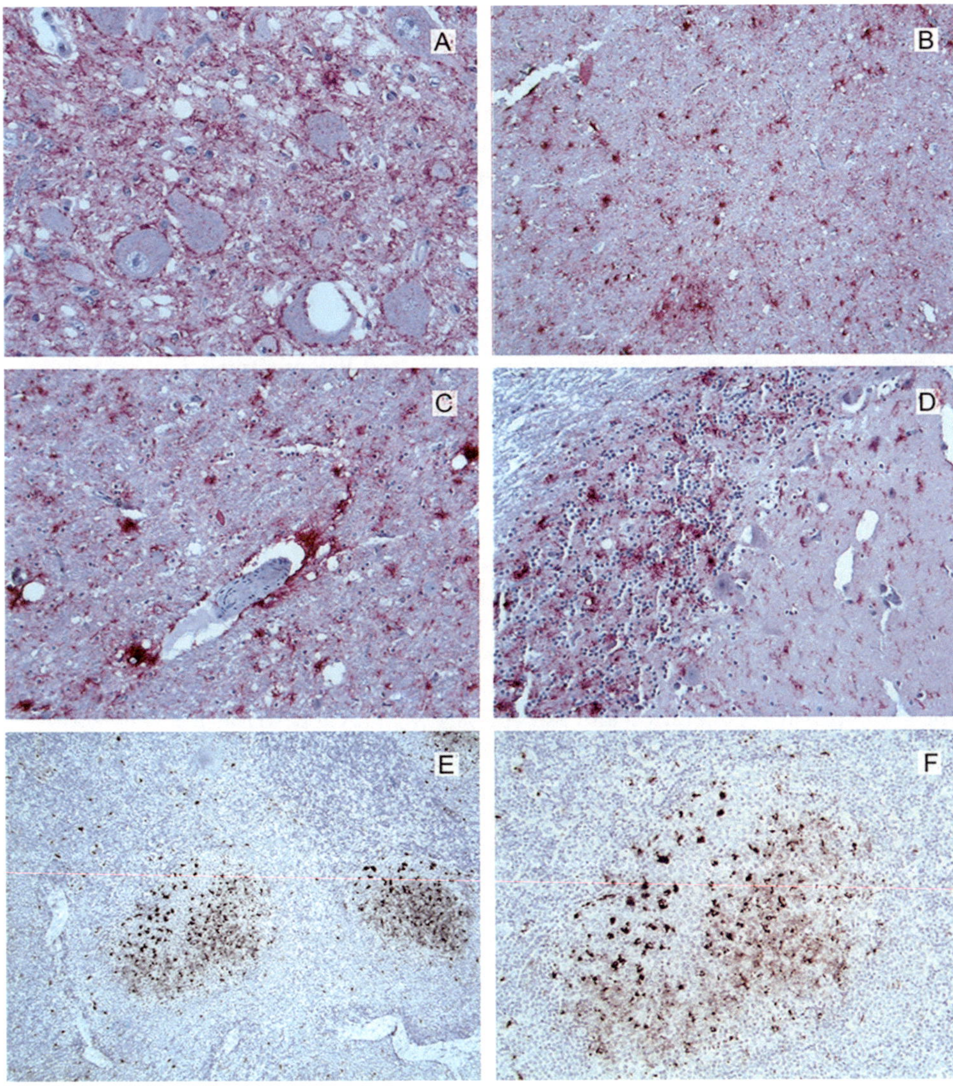

Figure 5.
PrPsc immunohistochemistry in CWD affected deer. (A) Perineuronal and granular deposits in neuropil at dorsal vagal nucleus (brainstem), X400; (B) Extracelullar plaque-like deposits (cerebral cortex), X200; (C) perivascular type (brainstem), X200; (D) distribution in both molecular and granular layers of the cerebellum, X200; (E) and (F) presence of PrPsc in several follicules in a lymph node, X100 and X200, respectivelly. (A–D) sections from CWD control samples kindly provided by professor Stefanie Czub (Canadian food inspection agency, National Center for animal diseases) in 2003; (E-F) sections from the CWD proficiency testing 2008 organized by the European reference laboratory for TSEs (APHA, Weybridge). (A-D) specific PrPsc signal was visualized with F99/97.6.1 monoclonal antibody (raised against bovine residues 220–225; VMRD Inc., Pullman, WA; 1:1250 dilution) StreptABC-alkaline phosphatase, new Fuchsin system (DAKO); (E and F) IHC with 2G11 monoclonal antibody (raised against ovine PrP peptide sequence 146-R154-R171–182; Institute Pourquier;1:200 dilution). All tissue sections were counterstained with Mayer's hematoxylin.

If histopathological examination cannot be carried out due to poor state of the sample and/or IHC results which are not conclusive, then Western immunoblotting is the remaining confirmation method available. This technique relies on detergent extraction followed by treatment with proteinase K to digest any PrP^{C} and detect only PrP^{SC} by a specific antibody presenting bands that correspond to proteins within a range of molecular mass from 17 kDa (unglycosylated PrP^{SC}) to 27 kD (diglycosylated PrP^{SC}) [238].

WB and IHC allow a comparison of labelling patterns with antibodies that recognize different epitopes of the protein and help to clarify specific proteinase K cleavage inherent of strains type. This approach forms the basis of the discriminatory testing of small ruminant isolates, to differentiate between isolates that can be classified as scrapie, and those that are considered BSE-like by looking at the lower molecular weight for the unglycosylated protein band and the ratio of the mono- and diglycosylated fragments, for example, between classical, atypical/Nor98, and CH1641 scrapie in sheep and goats and BSE and scrapie [246, 247].

The Norwegian atypical CWD (Nor-16CWD) also presented an unusual PrP^{sc} electrophoretic pattern distinguishable from previous CWD cases and from known ruminant prion diseases in Europe, with the possible exception of sheep CH1641. Transmission studies in several rodent models are ongoing to clarify if this different phenotype could reflect the presence of a new cervid prion strain in moose from Norway [31].

6.4 New diagnostic methods

Regarding diagnosis in vivo, more sensitive diagnostic methods are desirable. Using in vitro conversion such as protein misfolding cycling amplification (PMCA) or real-time quaking-induced conversion assay (RT-QuIC), CWD prions are detectable already at a preclinical stage in specimen that can be obtained antemortem by noninvasive methods, such as blood, urine, feces, or saliva.

During PMCA, the normal form of PrP (PrP^{c}) is converted into protease K-resistant PrP (PrP^{sc}) using small amounts of infectious PrP^{sc}. Continued recruitment and conversion of PrP^{c} by PrP^{sc} are accomplished by sonication in a process analogous to amplification of DNA by the polymerase chain reaction (PCR). This amplification process enhances detection sensitivity by several orders of magnitude as compared to WB and has been used to confirm the presence of CWD prions in muscle [248] and feces [249].

In the assay referred to as real-time quaking-induced conversion (RT-QuIC) [250], prion seeds are thought to induce recombinant PrP to adopt a β-sheet structure. Thioflavin T, added to the reaction, is incorporated into the growing amyloid causing an altered spectrofluorimetric emission pattern, which is monitored over time.

CWD still represents a challenge in TSE research for which transmission and dissemination remain unchecked. The tools of diagnostic available for identifying infected animals have steadily progressed over time from clinical and pathological descriptions to antibody–antigen-dependent immunoassays and more recently qualitative and quantitative prion amplification techniques. These tools have provided a deep understanding of disease pathogenesis and transmission and allowed animal health technicians to monitor the expanding geographical presence of CWD [251].

7. Surveillance and control of CWD

7.1 Surveillance

Of further concern is the fact that CWD is the only transmissible spongiform encephalopathy known to be expanding both geographically and in prevalence [5].

After the confirmation in 2016 of two cases of CWD in a wild reindeer (April) and a wild moose (May) in Norway (the first CWD cases in Europe), this country intensified its surveillance program for CWD in cervids and detected a number of other cases of CWD in reindeers and in moose. In the meantime, the European Commission (EC) requested the European Food Safety Agency (EFSA) to recommend surveillance activities and, if necessary, additional animal health risk-based measures to prevent the introduction of the disease and the spread into/within the European Union (EU), specifically Estonia, Finland, Iceland, Latvia, Lithuania, Norway, Poland, and Sweden [252]. Based on EFSA opinion, the EC places in force the Commission Regulation (EU) 2017/1972, amending Annexes I and III of Regulation (EC) No. 999/2001 as regards a surveillance program for chronic wasting disease in cervids in Estonia, Finland, Latvia, Lithuania, Poland, and Sweden and repealing Commission Decision 2007/182/EC (which provided requirements for a survey on CWD in cervids which was carried out from 2007 to 2010). According to this new regulation, the member states (MSs), which have a wild and/or farmed and/or semidomesticated population of moose band/or reindeer (aforementioned MSs), shall carry out a 3-year monitoring program for CWD from 1 January 2018 to 31 December 2020.

The 3-year CWD monitoring program shall cover animals over 12 months of age, estimated on the basis of dentition, obvious signs of maturity, or any other reliable information, of the following cervid species:

- Eurasian tundra reindeer (*Rangifer tarandus tarandus*)
- Finnish forest reindeer (*Rangifer tarandus fennicus*)
- Moose (*Alces alces*)
- Roe deer (*Capreolus capreolus*)
- White-tailed deer (*Odocoileus virginianus*)
- Red deer (*Cervus elaphus*)

According to some authors [252, 253], in order to avoid unbiased prevalence or incidence estimates and to obtain a representative set of animals, a random sampling is always recommended to design monitoring program. This classical approach may be possible to achieve for farmed cervids, but, for wildlife population, random sampling may not be possible to design as sampling frames are not available [252]. However, when disease detection is the first objective, a sampling targeting high-risk animals may be more efficient. That, in fact, was the case of the Norwegian monitoring program that detected the two CWD cases through sampling of "suspect" cervids [252].

Since the aims of the proposed surveillance system were to detect disease in countries where CWD has not yet been detected and to estimate prevalence in areas where disease has been detected, the 3-year surveillance system includes a two-stage sampling program (Commission Regulation (EU) 2017/1972):

- At the first stage, a random sampling is applied for wild/semidomesticated cervids [primary sampling (PSU)] corresponding to geographical areas containing cervid populations, whereas for farmed cervids they will correspond to farms.
- At the second stage, a convenience sampling is applied targeting high-risk animals of the selected species within PSU (**Figure 6**).

To date, according to Commission Regulation (EU) 2017/1972, member states shall carry out additional monitoring for TSEs in cervids based on a risk assessment which may take into account the detection of a TSE in cervids in the same or neighboring regions. Efforts to prevent the introduction of CWD should consider whether prior surveillance data are sufficient to assure that the disease has not already spread into the area of interest.

Based on the main provisions in the TSE regulation currently applicable to CWD, TSE-positive cases in cervids must be notified to the commission and the member states (Article 11 of Regulation (EC) No. 999/2001), and all parts of the body of a cervid positive for TSE must be sent to disposal as category 1 materials in accordance with the Animal By-Product Regulation 10 (Article 13.1. (a) of Regulation (EC) No. 999/2001).

7.2 TSE control

As it was previously referred (see Section 4), CWD is a very contagious form of TSE. Therefore, control in captive cervids is not easy and even more in the wildlife. According to Uehlinger and colleagues [254], the combination of direct and environmental transmission of CWD prions; the persistence of environmental infectious prion for at least for 2.5 years [255]; the absence of a rapid, practical, and reliable antemortem field tests for detecting preclinical CWD; and the nonavailability of licensed vaccines nor therapies together with the impossibility of quarantine suspicions populations in wild cervids make the control of this disease in wild populations very difficult.

The management of CWD in affected populations is a proven difficult task, based on population reduction in the case of wild deer or in quarantine and depopulation in commercially reared animals.

Measures to control the spread of the disease will depend on whether we are dealing with commercially exploited herds, confined to demarcated areas, or with wild populations of deer. In the first case, and similar to other infectious diseases and other livestock, programs to detect and eliminate positive animals may be used [94].

This task is not easy, however, as a shown by a 2018 study conducted on a private farm in Colorado where the use of disease control management through rectal biopsy diagnosis was not able to prevent the increase in prevalence suggesting the need to refine management criteria or antemortem diagnostic methods [5].

Thus, the science available to inform effective CWD management and control strategies remains relatively incomplete as it was concluded by different studies [254, 256] during their review on CWD control in North America. Nevertheless, it may be stated that approaches for CWD management generally fall into three categories in order to prevent the introduction of the disease and its spread [252]:

a. Prevention of entry in free regions (where CWD is assumed absent or is not believed to occur)

b. Containment procedures of a CWD focus to avoid disease spread

c. Control/suppression in an affected herd or population

Regarding preventive and containment strategies, it was referred that these measures tend to focus on regulations (e.g., bans on movements of live animals, carcasses, or specified risk materials) and efforts to prevent the introduction of CWD should consider whether prior surveillance data are sufficient to assure that the disease has not already spread into the area of interest [252].

Haley and collaborators [94] suggest that outer mechanisms of disease control could be useful like the development of simple antemortem test to use prior to animal movements among farms, the breeding for resistant genotypes, the installation of effective fences to prevent transmission between commercial

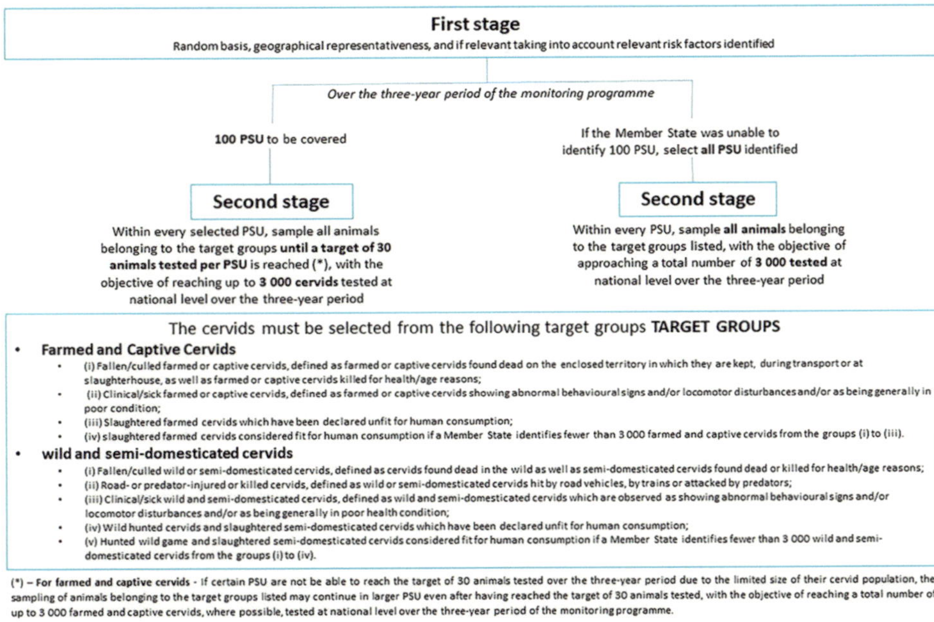

Figure 6.
Surveillance program for CWD in cervids. Schematic presentation of two-stage sampling approach of the 3-year surveillance programme for CWD in cervids in Estonia, Finland, Latvia, Lithuania, Poland and Sweden in cervids defined on commission regulation (EU) 2017/1972.

and wild populations, and equipment cleaning and quarantine prior to animal introduction.

Free-ranging deer population reduction is empirically effective in reducing a disease with direct contact transmission, but several studies have shown that in the case of CWD, the results are below expected [94, 257].

Capturing, testing, and recapturing positive free-ranging mule deer did not reduce prevalence in the female population [258].

Concerning the prevention of entry in free regions, according to Regulation (EC) No. 999/2001, in the EU, the feeding to cervids of proteins derived from animals is prohibited, with the exception of milk and milk products, eggs and egg products, hydrolyzed proteins from nonruminants or from ruminant hides and skins, and gelatine and collagen from nonruminants (Article 7 and Annex IV). Also, at import into the EU, an attestation is required for meat and meat products from wild and farmed cervids coming from the USA or Canada (Chapter F of Annex IX), confirming that the products:

- Exclude the offal and spinal cord
- Are derived from animals tested for CWD with negative results
- Are derived from animals which do not come from a herd (for farmed animals) or a region (for wild animals) where CWD has been confirmed or officially suspected

As a result of the detection of the first case in Norway, in Europe, in addition to the permanent measures as described in the aforementioned regulation, the Commission Implementing Decision (EU) 2016/1918 enforced temporary safeguard measures in relation to CWD (until December 31, 2017). According to this

DOI: http://dx.doi.org/10.5772/intechopen.85797

decision, the movement of live cervids from Norway into the European Union was prohibited with several derogations permitting its movements under certain circumstances (e.g., movements of live cervids from Norway to Sweden or Finland for direct slaughter; movements of live reindeer for seasonal grazing from Norway to some listed areas in Sweden and Finland).

According to the European Food Safety Agency [252], the derogations that lay down in this decision present a risk of introduction of CWD into the EU, since it was concluded that the most likely pathway of introduction of CWD into the EU is the movement of live cervids, either by deliberate transportation or by the movement of wild animals across the border of Norway to Sweden or Finland. Additionally, these authors present other measures to reduce the probability of introduction of CWD into the EU, namely, the use of natural cervid urine lures and awareness campaigns targeting both local Norwegian hunters and hunters visiting Norway from (and returning to) other countries regarding the personal protective equipment (PPE), disinfection, the safe dressing of carcasses, and the appropriate disposal of carcass trimmings.

In general, according to several authors [252, 254, 256], other measures can be implemented in order to contain and/or to control/suppress CWD in a region or country where the disease is present. Some of those measures may include:

- Reducing environmental contamination (e.g., interdict the use of dead cervids to feed wildlife nor to use as baits for hunting, unless they have been tested negative, prior to use).

- Reduction of animal-to-animal contact (e.g., ban artificial feeding, lick blocks or other congregating management procedures).

- Decrease population densities by definition of strategic plan for hunting management policies (e.g. increasing harvest permits and/or hunting season) and culling practices.

Regarding the last measure (decrease population), Uehlinger and colleagues [254] alert that:

- To date, the evidence is unclear that increased nonselective culling pressure has a beneficial effect on the spread or prevalence of CWD.

- Culling of wildlife is often unpopular. For that reason, any control program that includes culling of animals must take into account the public acceptance and hunters' attitudes and behaviors.

Since dispersion of infected deer may not be limited by political boundaries, regional and interstate/interprovincial cooperation and collaboration will be an important part of any successful management strategy. Also, it is always important to evaluate the costs in order to understand the feasibility of the control strategy proposals (when developing any intervention program) [254].

This review gathers known and recent features of this progressive and fatal neurodegenerative disease affecting cervid species, contributing for chronic wasting disease awareness among scientific community and stakeholders. Conscious that research in this field has a long way to go to answer many of the questions that remain open, we believe that this review will contribute to the understanding of the occurrence of prion diseases in wildlife.

Acknowledgements

We would like to thank Professor Jesus R. Requena (CIMUS Biomedical Research Institute and Department of Medical Sciences, University of Santiago de Compostela-IDIS, Spain) for his critical review in the description of the structure of PrP^{sc} in the Biology of the etiologic agent section.

Author details

Carla Neves Machado[1], Leonor Orge[1,2], Isabel Pires[2], Adelina Gama[2], Alexandra Esteves[2], Ana Paula Mendonça[1], Ana Matos[3], Anabela Alves[2], Carla Lima[1], Estela Bastos[4], Fernanda Seixas[2], Filipe Silva[2], João Carlos Silva[1], Luis Figueira[3], Madalena Vieira-Pinto[2], Maria De Lurdes Pinto[2], Nuno Gonçalves-Anjo[5], Paula Tavares[1], Paulo Carvalho[1], Roberto Sargo[2] and Maria Dos Anjos Pires[2*]

1 Pathology Laboratory, UEISPSA, National Institute for Agricultural and Veterinary Research (INIAV), Oeiras and Vila do Conde, Portugal

2 Animal and Veterinary Research Centre (CECAV), University of Trás-os-Montes and Alto Douro (UTAD), Vila Real, Portugal

3 Polytechnic Institute of Castelo Branco (IPCB), Castelo Branco, Portugal

4 Centre for the Research and Technology of Agro-Environmental and Biological Sciences (CITAB), University of Trás-os-Montes and Alto Douro (UTAD), Vila Real, Portugal

5 Genetics and Biotechnology Department, University of Trás-os-Montes and Alto Douro (UTAD), Vila Real, Portugal

*Address all correspondence to: apires@utad.pt

References

[1] Wood J, Lund L, Done S. The natural occurrence of scrapie in moufflon. The Veterinary Record. 1992;**130**:25-27

[2] Pitarch J, Raksa H, Arnal M, Revilla M, Martínez D, Fernández de Luco D, et al. Low sequence diversity of the prion protein gene (PRNP) in wild deer and goat species from Spain. Veterinary Research. 2018;**49**(1):33

[3] Peletto S, Perucchini M, Acin C, Dalgleish M, Reid H, Rasero R, et al. Genetic variability of the prion protein gene (PRNP) in wild ruminants from Italy and Scotland. Journal of Veterinary Science. 2009;**10**:115-120

[4] Hadjisterkotis E, editor The Cyprus Mouflon, a threatened species in a biodiversity "hotspot" area. In: Proceedings of the International Mouflon Symposium; Sopron, Hungary. 2001

[5] Haley NJ, Henderson DM, Wycoff S, Tennant J, Hoover EA, Love D, et al. Chronic wasting disease management in ranched elk using rectal biopsy testing. Prion. 2018;**12**(2):93-108

[6] Prusiner S. Biology and genetics of prions causing neurodegeneration. Annual Review of Genetics. 2013;**47**:601-623

[7] Riesner D. The Prion theory: Background and basic information. In: Rabenau HF, Ciantl J, Doherr HW, editors. Prions—A Challenge for Science, Medicine and Public Health System Contributions to Microbiology. Vol. 7. Basel: Karger; 2001. pp. 7-20

[8] Aguilar-Calvo P, García C, Espinosa JC, Andreoletti O, TJ M. Prion and prion-like diseases in animals. Virus Research. 2015;**207**:82-93

[9] Requena J, Wille H. The structure of the infectious prion protein Experimental data and molecular models. Prion. 2014;**8**:1-7

[10] Wille H, Requena J. The structure of PrPSc prions. Pathogens. 2018;**7**:20

[11] Wulf MA, Senatore A, Aguzzi A. The biological function of the cellular prion protein: An update. BMC Biology. 2017;**15**(1):34

[12] Dormont D. Prion diseases: Pathogenesis and public health concerns. FEBS Letters. 2002;**529**:17-21

[13] Huang Z, Prusiner S, Cohen F. Scrapie prions: A three-dimensional model of an infectious fragment. Folding and Design. 1996;**1**:13-19

[14] Riek R, Hornemann S, Wider G, Billeter M, Glockshuber R, Wüthrich K. NMR structure of the mouse prion protein domain PrP(121-231). Nature. 1996;**382**:180-182

[15] Downing D, Lazo N. Molecular modelling indicates that the pathological conformations of prion proteins might be beta-helical. The Biochemical Journal. 1999;**343**:453-460

[16] Wille H, Michelitsch M, Guenebaut V, Supattapone S, Serban A, Cohen F, et al. Structural studies of the scrapie prion protein by electron crystallography. Proceedings of the National Academy of Sciences of the United States of America. 2002;**99**:3563-3568

[17] Mornon J, Prat K, Dupuis F, Boisset N, Callebaut I. Structural features of prions explored by sequence analysis. II. A PrP(Sc) model. Cellular and Molecular Life Sciences. 2002;**59**:2144-2154

[18] DeMarco M, Daggett V. From conversion to aggregation: Protofibril formation of the prion protein.

Proceedings of the National Academy of Sciences of the United States of America. 2004;**101**:2293-2298

[19] Govaerts C, Wille H, Prusiner S, Cohen F. Evidence for assembly of prions with left-handed beta-helices into trimers. Proceedings of the National Academy of Sciences of the United States of America. 2004;**101**:8342-8347

[20] Stork M, Giese A, Kretzschmar H, Tavan P. Molecular dynamics simulations indicate a possible role of parallel beta-helices in seeded aggregation of poly-Gln. Biophysical Journal. 2005;**88**:2442-2451

[21] Yang S, Levine H, Onuchic J, Cox D. Structure of infectious prions: Stabilization by domain swapping. The FASEB Journal. 2005;**19**:1778-1782

[22] Langedijk J, Fuentes G, Boshuizen R, Bonvin A. Two-rung model of a left-handed beta-helix for prions explains species barrier and strain variation in transmissible spongiform encephalopathies. American Journal of Molecular Biology. 2006;**360**:907-920

[23] Cobb N, Sönnichsen F, McHaourab H, Surewicz W. Molecular architecture of human prion protein amyloid: A parallel, in-register beta-structure. Proceedings of the National Academy of Sciences of the United States of America. 2007;**104**:18946-18951

[24] Groveman B, Dolan M, Taubner L, Kraus A, Wickner R, Caughey B. Parallel in-register intermolecular β-sheet architectures for prion-seeded prion protein (PrP) amyloids. The Journal of Biological Chemistry. 2014;**289**:24129-24142

[25] Kunes K, Clark S, Cox D, Singh R. Left handed beta helix models for mammalian prion fibrils. Prion. 2008;**2**:81-90

[26] Vázquez-Fernández E, Vos M, Afanasyev P, Cebey L, Sevillano A, Vidal E, et al. The structural architecture of an infectious mammalian prion using electron cryomicroscopy. PLoS Pathogens. 2016;**8**:e1005835

[27] Angers R, Kang H, D N BS, Seward T, Mathiason C, et al. Prion strain mutation determined by prion protein conformational compatibility and primary structure. Science. 2010;**328**:1154-1158

[28] Benestad S, Telling G. Chronic wasting disease: An evolving prion disease of cervids. In: Pocchiari M, Manson J, editors. Handbook of Clinical Neurology, Human Prion Diseases. Vol. 153. Amsterdam: Elsevier BV; 2018. pp. 135-151

[29] Lee YH, Sohn HJ, Kim MJ, Kim HJ, Park KJ, Lee WY, et al. Experimental chronic wasting disease in wild type VM mice. The Journal of Veterinary Medical Science. 2013;**75**(8):1107-1110

[30] Benestad S, Mitchell G, Simmons M, Ytrehus B, Vikøren T. First case of chronic wasting disease in Europe in a Norwegian free-ranging reindeer. Veterinary Research. 2016;**47**:88

[31] Pirisinu L, Tran L, Chiappini B, Vanni I, Di Bari MA, Vaccari G, et al. Novel type of chronic wasting disease detected in moose (*Alces alces*), Norway. Emerging Infectious Diseases. 2018;**24**(12):2210-2218

[32] Martin S, Jeffrey M, Gonzalez L, Siso S, Reid HW, Steele P, et al. Immunohistochemical and biochemical characteristics of BSE and CWD in experimentally infected European red deer (*Cervus elaphus elaphus*). BMC Veterinary Research. 2009;**5**:26

[33] Kurt TD, Jiang L, Fernández-Borges N, Bett C, Liu J, Yang T, et al. Human prion protein sequence elements impede cross-species chronic wasting disease

transmission. The Journal of Clinical Investigation. 2015;**125**(4):1485-1496

[34] Crick F. Central dogma of molecular biology. Nature. 1970;**227**(5258): 561-563

[35] Manning KS, Cooper TA. The roles of RNA processing in translating genotype to phenotype. Nature Reviews. Molecular Cell Biology. 2017;**18**(2):102-114

[36] Prusiner SB. Molecular structure, biology, and genetics of prions. In: Maramorosch K, Murphy FA, Shatkin AJ, editors. Advances in Virus Research. Vol. 35. Cambridge, Massachusetts: Academic Press; 1988. pp. 83-136

[37] Chakravarty AK, Jarosz DF. More than just a phase: Prions at the crossroads of epigenetic inheritance and evolutionary change. Journal of Molecular Biology. 2018;**430**(23):4607-4618

[38] Baskakov IV, Breydo L. Converting the prion protein: What makes the protein infectious. Biochimica et Biophysica Acta (BBA)—Molecular Basis of Disease. 2007;**1772**(6):692-703

[39] Fujun F, Lanhui L, Xianglong L, Xiangyun L, Rongyan Z. Evolution and differentiation of the prion protein gene (PRNP) among species. The Journal of Heredity. 2008;**99**(6):647-652

[40] Robinson SJ, Samuel MD, O'Rourke KI, Johnson CJ. The role of genetics in chronic wasting disease of North American cervids. Prion. 2012;**6**(2):153-162

[41] O'Rourke KI, Baszler TV, Miller JM, Spraker TR, Sadler-Riggleman I, Knowles DP. Monoclonal antibody F89/160.1.5 defines a conserved epitope on the ruminant prion protein. Journal of Clinical Microbiology. 1998;**36**(6):1750-1755

[42] Owen F, Poulter M, Collinge J, Crow TJ. Codon 129 changes in the prion protein gene in Caucasians. The American Journal of Human Genetics. 1990;**46**(6):1215-1216

[43] Collinge J, Palmer MS, Dryden AJ. Genetic predisposition to iatrogenic Creutzfeldt-Jakob disease. The Lancet. 1991;**337**(8755):1441-1442

[44] Palmer MS, Dryden AJ, Hughes JT, Collinge J. Homozygous prion protein genotype predisposes to sporadic Creutzfeldt-Jakob disease. Nature. 1991;**352**(6333):340-342

[45] O'Rourke KI, Besser TE, Miller MW, Cline TF, Spraker TR, Jenny AL, et al. PrP genotypes of captive and free-ranging Rocky Mountain elk (*Cervus elaphus* nelsoni) with chronic wasting disease. The Journal of General Virology. 1999;**80**(Pt 10):2765-2769

[46] O'Rourke KI, Spraker TR, Zhuang D, Greenlee JJ, Gidlewski TE, Hamir AN. Elk with a long incubation prion disease phenotype have a unique PrPd profile. NeuroReport. 2007;**18**(18):1935-1938

[47] Goldmann W. PrP genetics in ruminant transmissible spongiform encephalopathies. Journal of Veterinary Research. 2008;**39**(4):30

[48] O'Rourke KI, Spraker TR, Hamburg LK, Besser TE, Brayton KA, Knowles DP. Polymorphisms in the prion precursor functional gene but not the pseudogene are associated with susceptibility to chronic wasting disease in white-tailed deer. The Journal of General Virology. 2004;**85**(Pt 5):1339-1346

[49] Williams ES. Chronic wasting disease. Journal of Veterinary Pathology. 2005;**42**:530-549

[50] Huson HJ, Happ GM. Polymorphisms of the prion protein

gene (PRNP) in Alaskan moose (*Alces alces gigas*). Animal Genetics. 2006;**37**(4):425-426

[51] Jewell JE, Conner MM, Wolfe LL, Miller MW, Williams ES. Low frequency of PrP genotype 225SF among free-ranging mule deer (*Odocoileus hemionus*) with chronic wasting disease. Journal of General Virology. 2005;**86**(8):2127-2134

[52] Wilson GA, Nakada SM, Bollinger TK, Pybus MJ, Merrill EH, Coltman DW. Polymorphisms at the PRNP gene influence susceptibility to chronic wasting disease in two species of deer (*Odocoileus* Spp.) in western Canada. Journal of Toxicology and Environmental Health. Part A. 2009;**72**(17-18):1025-1029

[53] Johnson C, Johnson J, Vanderloo JP, Keane D, Aiken JM, McKenzie D. Prion protein polymorphisms in white-tailed deer influence susceptibility to chronic wasting disease. The Journal of General Virology. 2006;**87**(Pt 7):2109-2114

[54] Jeong HJ, Lee JB, Park SY, Song CS, Kim BS, Rho JR, et al. Identification of single-nucleotide polymorphisms of the prion protein gene in sika deer (*Cervus nippon* laiouanus). Journal of Veterinary Science. 2007;**8**(3):299-301

[55] Happ G, Huson H, Beckmen K, Kennedy L. Prion protein genes in caribou from Alaska. Journal of Wildlife Diseases. 2007;**43**(2):224-228

[56] Perucchini M, Griffin K, Miller MW, Goldmann W. PrP genotypes of free-ranging wapiti (*Cervus elaphus* nelsoni) with chronic wasting disease. The Journal of General Virology. 2008;**89**(Pt 5):1324-1328

[57] White SN, Spraker TR, Reynolds JO, O'Rourke KI. Association analysis of PRNP gene region with chronic wasting disease in Rocky Mountain elk. BMC Research Notes. 2010;**3**:314

[58] Heaton MP, Leymaster KA, Freking BA, Hawk DA, Smith TPL, Keele JW, et al. Prion gene sequence variation within diverse groups of U.S. sheep, beef cattle, and deer. Mammalian Genome. 2003;**14**(11):765-777

[59] Meng LP, Zhao DM, Liu HX, Yang JM, Ning ZY, Wu CD, et al. Polymorphisms of the prion protein gene (PRNP) in Chinese domestic sika deer (*Cervus nippon hortulorum*). Animal Genetics. 2005;**36**(3):266-267

[60] Williams ES, Young S. Chronic wasting disease of captive mule deer: A spongiform encephalopathy. Journal of Wildlife Diseases. 1980;**16**(1):89-98

[61] Guiroy DC, Williams ES, Yanagihara R, Gajdusek DC. Topographic distribution of scrapie amyloid-immunoreactive plaques in chronic wasting disease in captive mule deer (*Odocoileus hemionus hemionus*). Acta Neuropathologica. 1991;**81**(5):475-478

[62] Guiroy DC, Williams ES, Yanagihara R, Gajdusek DC. Immunolocalization of scrapie amyloid (PrP27-30) in chronic wasting disease of Rocky Mountain elk and hybrids of captive mule deer and white-tailed deer. Neuroscience Letters. 1991;**126**(2):195-198

[63] Williams ES, Young S. Spongiform encephalopathy of Rocky Mountain elk. Journal of Wildlife Diseases. 1982;**18**(4):465-471

[64] Williams ES, Young S. Spongiform encephalopathies in Cervidae. Revue Scientifique et Technique. 1992;**11**(2):551-567

[65] Williams ES. Chronic wasting disease. Veterinary Pathology. 2005;**42**(5):530-549

[66] Benestad SL, Telling GC. Chronic wasting disease: An evolving prion disease of cervids. Handbook of Clinical Neurology. 2018;**153**:135-151

[67] Williams ES, Miller MW. Chronic wasting disease in deer and elk in North America. Revue Scientifique et Technique. 2002;**21**(2):305-316

[68] Haley NJ, Henderson DM, Wycoff S, Tennant J, Hoover EA, Love D, et al. Chronic wasting disease management in ranched elk using rectal biopsy testing. Prion. 2018;**12**(2):93-108

[69] United States Geological Survey (USGS). Expanding Distribution of Chronic Wasting Disease. https://www.usgs.gov/centers/nwhc/science/expanding-distribution-chronic-wasting-disease?qt-science_center_objects=0#qt-science_center_objects (accessed 25 April 2019)

[70] Centers for Disease Control and Prevention: CDC 24/7: Saving Lives, Protecting People TM. Chronic Wasting Disease: Occurrence. https://www.cdc.gov/prions/cwd/occurrence.html (accessed 25 April 2019)

[71] Kahn S, Dube C, Bates L, Balachandran A. Chronic wasting disease in Canada: Part 1. The Canadian Veterinary Journal. 2004;**45**(5):397-404

[72] Ball K. Chronic wasting disease in a Rocky Mountain elk. The Canadian Veterinary Journal. 2002;**43**(11):880-882

[73] Dube C, Mehren KG, Barker IK, Peart BL, Balachandran A. Retrospective investigation of chronic wasting disease of cervids at the Toronto Zoo, 1973-2003. The Canadian Veterinary Journal. 2006;**47**(12):1185-1193

[74] Lapointe JM, Leclair D, Mesher C, Balachandran A. Screening for chronic wasting disease in caribou in northern Quebec. The Canadian Veterinary Journal. 2002;**43**(11):886-887

[75] Canadian Food Inspection Agency. Chronic wasting disease (CWD) of deer and elk. http://www.inspection.gc.ca/animals/terrestrial-animals/diseases/reportable/cwd/eng/1330143462380/1330143991594 (accessed 25 April 2019)

[76] Sohn HJ, Kim JH, Choi KS, Nah JJ, Joo YS, Jean YH, et al. A case of chronic wasting disease in an elk imported to Korea from Canada. The Journal of Veterinary Medical Science. 2002;**64**(9):855-858

[77] Ricci A, Allende A, Bolton D, Chemaly M, Davies R, Fernández Escámez PS, et al. Scientific opinion on chronic wasting disease (II). The European Food Safety Authority (EFSA) Journal. 2018;**16**(1):e05132

[78] Gale P, Roberts H. Update on chronic wasting disease in europe. department for environment. Food and Rural Affairs Animal & Plant Health Agency Monitoring VSPAT-ID. 2018;**2018**:25

[79] Joly DO, Ribic CA, Langenberg JA, Beheler K, Batha CA, Dhuey BJ, et al. Chronic wasting disease in free-ranging Wisconsin white-tailed deer. Emerging Infectious Diseases. 2003;**9**(5):599-601

[80] Williams ES, Young S. Neuropathology of chronic wasting disease of mule deer (*Odocoileus hemionus*) and elk (*Cervus elaphus* nelsoni). Veterinary Pathology. 1993;**30**(1):36-45

[81] Baeten LA, Powers BE, Jewell JE, Spraker TR, Miller MW. A natural case of chronic wasting disease in a free-ranging moose (*Alces alces shirasi*). Journal of Wildlife Diseases. 2007;**43**(2):309-314

[82] Race B, Meade-White KD, Phillips K, Striebel J, Race R, Chesebro B. Chronic wasting disease agents in nonhuman primates. Emerging Infectious Diseases. 2014;**20**(5):833-837

[83] Race B, Meade-White KD, Miller MW, Barbian KD, Rubenstein R, LaFauci G, et al. Susceptibilities of nonhuman primates to chronic wasting disease. Emerging Infectious Diseases. 2009;**15**(9):1366-1376

[84] Heisey DM, Mickelsen NA, Schneider JR, Johnson CJ, Johnson CJ, Langenberg JA, et al. Chronic wasting disease (CWD) susceptibility of several North American rodents that are sympatric with cervid CWD epidemics. Journal of Virology. 2010;**84**(1):210-215

[85] Di Bari MA, Nonno R, Castilla J, D'Agostino C, Pirisinu L, Riccardi G, et al. Chronic wasting disease in bank voles: Characterisation of the shortest incubation time model for prion diseases. PLoS Pathogens. 2013;**9**(3):e1003219

[86] Raymond GJ, Raymond LD, Meade-White KD, Hughson AG, Favara C, Gardner D, et al. Transmission and adaptation of chronic wasting disease to hamsters and transgenic mice: Evidence for strains. Journal of Virology. 2007;**81**(8):4305-4314

[87] Perrott MR, Sigurdson CJ, Mason GL, Hoover EA. Evidence for distinct chronic wasting disease (CWD) strains in experimental CWD in ferrets. The Journal of General Virology. 2012;**93** (Pt 1):212-221

[88] Hamir AN, Miller JM, Cutlip RC, Stack MJ, Chaplin MJ, Jenny AL, et al. Experimental inoculation of scrapie and chronic wasting disease agents in raccoons (*Procyon lotor*). The Veterinary Record. 2003;**153**(4):121-123

[89] Mathiason CK, Nalls AV, Seelig DM, Kraft SL, Carnes K, Anderson KR, et al. Susceptibility of domestic cats to chronic wasting disease. Journal of Virology. 2013;**87**(4):1947-1956

[90] Moore SJ, West Greenlee MH, Kondru N, Manne S, Smith JD, Kunkle RA, et al. Experimental transmission of the chronic wasting disease agent to swine after oral or intracranial inoculation. Journal of Virology. 2017;**91**(19)

[91] Hamir AN, Kunkle RA, Cutlip RC, Miller JM, Williams ES, Richt JA. Transmission of chronic wasting disease of mule deer to Suffolk sheep following intracerebral inoculation. Journal of Veterinary Diagnostic Investigation. 2006;**18**(6):558-565

[92] Hamir AN, Greenlee JJ, Nicholson EM, Kunkle RA, Richt JA, Miller JM, et al. Experimental transmission of chronic wasting disease (CWD) from elk and white-tailed deer to fallow deer by intracerebral route: Final report. Canadian Journal of Veterinary Research. 2011;**75**(2):152-156

[93] Nalls AV, McNulty E, Powers J, Seelig DM, Hoover C, Haley NJ, et al. Mother to offspring transmission of chronic wasting disease in reeves' muntjac deer. PLoS One. 2013;**8**(8):e71844

[94] Haley NJ, Hoover EA. Chronic wasting disease of cervids: Current knowledge and future perspectives. Annual Review of Animal Biosciences. 2015;**3**:305-325

[95] DeVivo MT, Edmunds DR, Kauffman MJ, Schumaker BA, Binfet J, Kreeger TJ, et al. Endemic chronic wasting disease causes mule deer population decline in Wyoming. PLoS One. 2017;**12**(10):e0186512

[96] Monello RJ, Powers JG, Hobbs NT, Spraker TR, O'Rourke KI, Wild MA. Efficacy of antemortem rectal biopsies to diagnose and estimate prevalence of chronic wasting disease in free-ranging cow elk (*Cervus elaphus* nelsoni). Journal of Wildlife Diseases. 2013;**49**(2):270-278

[97] Keane DP, Barr DJ, Bochsler PN, Hall SM, Gidlewski T, O'Rourke KI, et al. Chronic wasting disease in a

Wisconsin white-tailed deer farm. Journal of Veterinary Diagnostic Investigation. 2008;**20**(5):698-703

[98] Clements GM, Hygnstrom SE, Gilsdorf JM, Baasch DM, Clements MJ, Vercauteren KC. Movements of white-tailed deer in Riparian Habitat: Implications for infectious diseases. Journal of Wildlife Management. 2011;**75**(6):1436-1442

[99] Houston F, Andréoletti O. Chapter 25—The zoonotic potential of animal prion diseases. In: Pocchiari M, Manson J, editors. Handbook of Clinical Neurology, Human Prion Diseases. Vol. 153. Amsterdam: Elsevier BV; 2018. pp. 447-462

[100] Waddell L, Greig J, Mascarenhas M, Otten A, Corrin T, Hierlihy K. Current evidence on the transmissibility of chronic wasting disease prions to humans-A systematic review. Transboundary and Emerging Diseases. 2018;**65**(1):37-49

[101] Haley NJ, Siepker C, Walter WD, Thomsen BV, Greenlee JJ, Lehmkuhl AD, et al. Antemortem detection of chronic wasting disease prions in nasal brush collections and rectal biopsy specimens from white-tailed deer by real-time quaking-induced conversion. Journal of Clinical Microbiology. 2016;**54**(4):1108-1116

[102] Gough KC, Maddison BC. Prion transmission: Prion excretion and occurrence in the environment. Prion. 2010;**4**(4):275-282

[103] Race B, Meade-White K, Race R, Chesebro B. Prion infectivity in fat of deer with chronic wasting disease. Journal of Virology. 2009;**83**(18):9608-9610

[104] Saunders SE, Bartelt-Hunt SL, Bartz JC. Occurrence, transmission, and zoonotic potential of chronic wasting disease. Emerging Infectious Diseases. 2012;**18**(3):369-376

[105] VerCauteren KC, Pilon JL, Nash PB, Phillips GE, Fischer JW. Prion remains infectious after passage through digestive system of American crows (*Corvus brachyrhynchos*). PLoS One. 2012;**7**(10):e45774

[106] Fischer JW, Phillips GE, Nichols TA, Vercauteren KC. Could avian scavengers translocate infectious prions to disease-free areas initiating new foci of chronic wasting disease? Prion. 2013;**7**(4):263-266

[107] Nichols TA, Fischer JW, Spraker TR, Kong Q, VerCauteren KC. CWD prions remain infectious after passage through the digestive system of coyotes (*Canis latrans*). Prion. 2015;**9**(5):367-375

[108] Williams ES, Young S. Spongiform encephalopathies in Cervidae. Journal of Revue Scientifique et Technique. 1992;**11**(2):551-567

[109] Béringue V, Vilotte JL, Laude H. Prion agent diversity and species barrier. Veterinary Research. 2008;**39**(4):47

[110] Béringue V, Herzog L, Jaumain E, Reine F, Sibille P, Le Dur A, et al. Facilitated cross-species transmission of prions in extraneural tissue. Science. 2012;**335**(6067):472-475

[111] Crowell J, Hughson A, Caughey B, Bessen RA. Host determinants of prion strain diversity independent of prion protein genotype. Journal of Virology. 2015;**89**(20):10427-10441

[112] Barria MA, Telling GC, Gambetti P, Mastrianni JA, Soto C. Generation of a new form of human PrP(Sc) in vitro by interspecies transmission from cervid prions. The Journal of Biological Chemistry. 2011;**286**(9):7490-7495

[113] Bartz JC, Marsh RF, McKenzie DI, Aiken JM. The host range of chronic wasting disease is altered on passage in ferrets. Virology. 1998;**251**(2):297-301

[114] Harrathi C, Fernández-Borges N, Eraña H, Elezgarai SR, Venegas V, Charco JM, et al. Insights into the bidirectional properties of the sheep–deer prion transmission barrier. Molecular Neurobiology. 2018. DOI: 10.1007/s12035-018-1443-8

[115] Browning SR, Mason GL, Seward T, Green M, Eliason GA, Mathiason C, et al. Transmission of prions from mule deer and elk with chronic wasting disease to transgenic mice expressing cervid PrP. Journal of Virology. 2004;**78**(23):13345-13350

[116] Kong Q, Huang S, Zou W, Vanegas D, Wang M, Wu D, et al. Chronic wasting disease of elk: Transmissibility to humans examined by transgenic mouse models. The Journal of Neuroscience. 2005;**25**(35):7944-7949

[117] LaFauci G, Carp RI, Meeker HC, Ye X, Kim JI, Natelli M, et al. Passage of chronic wasting disease prion into transgenic mice expressing Rocky Mountain elk (*Cervus elaphus* nelsoni) PrPC. The Journal of General Virology. 2006;**87**(Pt 12):3773-3780

[118] Tamgüney G, Giles K, Bouzamondo-Bernstein E, Bosque PJ, Miller MW, Safar J, et al. Transmission of elk and deer prions to transgenic mice. Journal of Virology. 2006;**80**(18):9104-9114

[119] Madsen-Bouterse SA, Schneider DA, Zhuang D, Dassanayake RP, Balachandran A, Mitchell GB, et al. Primary transmission of chronic wasting disease versus scrapie prions from small ruminants to transgenic mice expressing ovine or cervid prion protein. The Journal of General Virology. 2016;**97**(9):2451-2460

[120] Scott M, Foster D, Mirenda C, Serban D, Coufal F, Wälchli M, et al. Transgenic mice expressing hamster prion protein produce species-specific scrapie infectivity and amyloid plaques. Cell. 1989;**59**(5):847-857

[121] Scott M, Groth D, Foster D, Torchia M, Yang S-L, DeArmond SJ, et al. Propagation of prions with artificial properties in transgenic mice expressing chimeric PrP genes. Cell. 1993;**73**(5):979-988

[122] Hara H, Hanada K, Ki H. Species-barrier phenomenon in prion transmissibility from a viewpoint of protein science. The Journal of Biochemistry. 2013;**153**(2):139-145

[123] Vanik DL, Surewicz KA, Surewicz WK. Molecular basis of barriers for interspecies transmissibility of mammalian prions. Molecular Cell. 2004;**14**(1):139-145

[124] Sandberg MK, Al-Doujaily H, Sigurdson CJ, Glatzel M, O'Malley C, Powell C, et al. Chronic wasting disease prions are not transmissible to transgenic mice overexpressing human prion protein. The Journal of General Virology. 2010;**91**(Pt 10):2651-2657

[125] Ironside JW, Head MW. Exploring the zoonotic potential of animal prion diseases AU—Barria, Marcelo A. Prion. 2014;**8**(1):85-91

[126] Lockey R, Beck KE, Simmons MM, Sallis RE, Spiropoulos J. Ovine PrP genotype is linked with lesion profile and immunohistochemistry patterns after primary transmission of classical scrapie to wild-type mice. Journal of Neuropathology and Experimental Neurology. 2010;**69**(5):483-497

[127] Beck KE, Sallis RE, Lockey R, Vickery CM, Béringue V, Laude H, et al. Use of murine bioassay to resolve ovine transmissible spongiform encephalopathy cases showing a bovine spongiform encephalopathy molecular profile. Brain pathology (Zurich, Switzerland). 2012;**22**(3):265-279

[128] Beck KE, Vickery CM, Lockey R, Holder T, Thorne L, Terry LA, et al. The interpretation of disease phenotypes

to identify TSE strains following murine bioassay: Characterisation of classical scrapie. Veterinary Research. 2012;**43**(1):77

[129] Bruce ME. TSE strain variation: An investigation into prion disease diversity. British Medical Bulletin. 2003;**66**(1):99-108

[130] Bruce ME, Boyle A, Cousens S, McConnell I, Foster J, Goldmann W, et al. Strain characterization of natural sheep scrapie and comparison with BSE. The Journal of General Virology. 2002;**83**:695-704

[131] Bencsik A, Vulin J, Lakhdar L, Baron T, Spiropoulos J, Beck KE. Selection of distinct strain phenotypes in mice infected by ovine natural scrapie isolates similar to CH1641 experimental scrapie. Journal of Neuropathology and Experimental Neurology. 2012;**71**(2):140-147

[132] Cordier C, Bencsik A, Philippe S, B_etemps D, Ronzon F, Calavas D, et al. Transmission and characterization of bovine spongiform encephalopathy sources in two ovine transgenic mouse lines (TgOvPrP4 and TgOvPrP59). Journal of General Virology. 2006;**87**:3763-3771

[133] Crozet C, Flamant F, Bencsik A, Aubert D, Samarut J, Baron T. Efficient transmission of two different sheep scrapie isolates in transgenic mice expressing the ovine PrP gene. Journal of Virology. 2001;**75**(11):5328-5334

[134] Vilotte JL, Soulier S, Essalmani R, Stinnakre MG, Vaiman D, Lepourry L, et al. Markedly increased susceptibility to natural sheep scrapie of transgenic mice expressing ovine prp. Journal of Virology. 2001;**75**(13):5977-5984

[135] Mitchell G, Yogasingam N, Walther I, Balachandran A. Experimental transmission of chronic wasting disease to sheep and goats—P70. Prion 2015 Poster Abstracts Y1-2015/04/24. Prion. 2015;**9**(suppl 1):S11-S99

[136] Luers L, Bannach O, Stöhr J, Wördehoff MM, Wolff M, Nagel-Steger L, et al. Seeded fibrillation as molecular basis of the species barrier in human prion diseases. PLoS One. 2013;**8**(8):e72623-e

[137] Prinz M, Huber G, Macpherson AJ, Heppner FL, Glatzel M, Eugster HP, et al. Oral prion infection requires normal numbers of Peyer's patches but not of enteric lymphocytes. The American Journal of Pathology. 2003;**162**(4):1103-1111

[138] Glaysher BR, Mabbott NA. Role of the GALT in scrapie agent neuroinvasion from the intestine. Journal of Immunology. 2007;**178**(6):3757-3766

[139] Donaldson DS, Else KJ, Mabbott NA. The Gut-associated lymphoid tissues in the small intestine, not the large intestine, play a major role in oral prion disease pathogenesis. Journal of Virology. 2015;**89**(18):9532-9547

[140] Donaldson DS, Mabbott NA. The influence of the commensal and pathogenic gut microbiota on prion disease pathogenesis. The Journal of General Virology. 2016;**97**(8):1725-1738

[141] Mabbott NA, Donaldson DS, Ohno H, Williams IR, Mahajan A. Microfold (M) cells: Important immunosurveillance posts in the intestinal epithelium. Mucosal Immunology. 2013;**6**(4):666-677

[142] Wyckoff AC, Kane S, Lockwood K, Seligman J, Michel B, Hill D, et al. Clay components in soil dictate environmental stability and bioavailability of cervid prions in mice. Frontiers in Microbiology. 2016;**7**:1885

[143] Kujala P, Raymond CR, Romeijn M, Godsave SF, van Kasteren SI,

Wille H, et al. Prion uptake in the gut: Identification of the first uptake and replication sites. PLoS Pathogens. 2011;7(12):e1002449

[144] Mabbott NA, Alibhai JD, Manson J. The role of the immune system in prion infection. Handbook of Clinical Neurology. 2018;**153**:85-107

[145] Kitamoto T, Muramoto T, Mohri S, Doh-Ura K, Tateishi J. Abnormal isoform of prion protein accumulates in follicular dendritic cells in mice with Creutzfeldt-Jakob disease. Journal of Virology. 1991;**65**(11):6292-6295

[146] McBride PA, Eikelenboom P, Kraal G, Fraser H, Bruce ME. PrP protein is associated with follicular dendritic cells of spleens and lymph nodes in uninfected and scrapie-infected mice. The Journal of Pathology. 1992;**168**(4):413-418

[147] Hilton DA, Fathers E, Edwards P, Ironside JW, Zajicek J. Prion immunoreactivity in appendix before clinical onset of variant Creutzfeldt-Jakob disease. Lancet. 1998;**352**(9129):703-704

[148] Sigurdson CJ, Williams ES, Miller MW, Spraker TR, O'Rourke KI, Hoover EA. Oral transmission and early lymphoid tropism of chronic wasting disease PrPres in mule deer fawns (*Odocoileus hemionus*). The Journal of General Virology. 1999;**80** (Pt 10):2757-2764

[149] Andreoletti O, Berthon P, Marc D, Sarradin P, Grosclaude J, van Keulen L, et al. Early accumulation of PrP(Sc) in gut-associated lymphoid and nervous tissues of susceptible sheep from a Romanov flock with natural scrapie. The Journal of General Virology. 2000;**81**(Pt 12):3115-3126

[150] Mitchell DA, Kirby L, Paulin SM, Villiers CL, Sim RB. Prion protein activates and fixes complement directly via the classical pathway: Implications for the mechanism of scrapie agent propagation in lymphoid tissue. Molecular Immunology. 2007;**44**(11):2997-3004

[151] Sim RB, Kishore U, Villiers CL, Marche PN, Mitchell DA. C1q binding and complement activation by prions and amyloids. Immunobiology. 2007;**212**(4-5):355-362

[152] Klein MA, Kaeser PS, Schwarz P, Weyd H, Xenarios I, Zinkernagel RM, et al. Complement facilitates early prion pathogenesis. Nature Medicine. 2001;**7**(4):488-492

[153] Mabbott NA, Bruce ME, Botto M, Walport MJ, Pepys MB. Temporary depletion of complement component C3 or genetic deficiency of C1q significantly delays onset of scrapie. Nature Medicine. 2001;7(4):485-487

[154] Zabel M, Heikenwälder M, Prinz M, Arrighi I, Schwarz P, Kranich J, et al. Stromal complement receptor CD21/35 facilitates lymphoid prion colonization and pathogenesis. The Journal of Immunology. 2007;**179**(9):6144-6152

[155] Michel B, Ferguson A, Johnson T, Bender H, Meyerett-Reid C, Pulford B, et al. Genetic depletion of complement receptors CD21/35 prevents terminal prion disease in a mouse model of chronic wasting disease. Journal of Immunology (Baltimore, MD: 1950). 2012;**189**(9):4520-4527

[156] Michel B, Ferguson A, Johnson T, Bender H, Meyerett-Reid C, Wyckoff AC, et al. Complement protein C3 exacerbates prion disease in a mouse model of chronic wasting disease. International Immunology. 2013;**25**(12):697-702

[157] Shortman K, Liu YJ. Mouse and human dendritic cell subtypes. Nature Reviews. Immunology. 2002;**2**(3):151-161

[158] Mabbott NA, Kenneth Baillie J, Kobayashi A, Donaldson DS, Ohmori H, Yoon S-O, et al. Expression of mesenchyme-specific gene signatures by follicular dendritic cells: Insights from the meta-analysis of microarray data from multiple mouse cell populations. Immunology. 2011;**133**(4):482-498

[159] McGovern G, Mabbott N, Jeffrey M. Scrapie affects the maturation cycle and immune complex trapping by follicular dendritic cells in mice. PLoS One. 2009;**4**(12):e8186-e

[160] O'Connor T, Frei N, Sponarova J, Schwarz P, Heikenwalder M, Aguzzi A. Lymphotoxin, but not TNF, is required for prion invasion of lymph nodes. PLoS Pathogens. 2012;**8**(8):e1002867

[161] Fraser H, Brown KL, Stewart K, McConnell I, McBride P, Williams A. Replication of scrapie in spleens of SCID mice follows reconstitution with wild-type mouse bone marrow. The Journal of General Virology. 1996;77 (Pt 8):1935-1940

[162] Klein MA, Frigg R, Flechsig E, Raeber AJ, Kalinke U, Bluethmann H, et al. A crucial role for B cells in neuroinvasive scrapie. Nature. 1997;**390**(6661):687-690

[163] Mohan J, Brown KL, Farquhar CF, Bruce ME, Mabbott NA. Scrapie transmission following exposure through the skin is dependent on follicular dendritic cells in lymphoid tissues. Journal of Dermatological Science. 2004;**35**(2):101-111

[164] McFarlin DE, Raff MC, Simpson E, Nehlsen SH. Scrapie in immunologically deficient mice. Nature. 1971;**233**(5318):336

[165] Fraser H, Dickinson AG. Studies of the lymphoreticular system in the pathogenesis of scrapie: The role of spleen and thymus. Journal of Comparative Pathology. 1978;**88**(4):563-573

[166] Raeber AJ, Sailer A, Hegyi I, Klein MA, Rülicke T, Fischer M, et al. Ectopic expression of prion protein (PrP) in T lymphocytes or hepatocytes of PrP knockout mice is insufficient to sustain prion replication. Proceedings of the National Academy of Sciences. 1999;**96**(7):3987

[167] Montrasio F, Cozzio A, Flechsig E, Rossi D, Klein MA, Rülicke T, et al. B lymphocyte-restricted expression of prion protein does not enable prion replication in prion protein knockout mice. Proceedings of the National Academy of Sciences. 2001;**98**(7):4034

[168] Mackay F, Browning JL. Turning off follicular dendritic cells. Nature. 1998;**395**(6697):26-27

[169] Schmidt TH, Bannard O, Gray EE, Cyster JG. CXCR4 promotes B cell egress from Peyer's patches. The Journal of Experimental Medicine. 2013;**210**(6):1099-1107

[170] Suzuki K, Grigorova I, Phan TG, Kelly LM, Cyster JG. Visualizing B cell capture of cognate antigen from follicular dendritic cells. The Journal of Experimental Medicine. 2009;**206**(7):1485-1493

[171] Mok SW, Proia RL, Brinkmann V, Mabbott NA. B cell-specific S1PR1 deficiency blocks prion dissemination between secondary lymphoid organs. Journal of Immunology. 2012;**188**(10):5032-5040

[172] Edwards JC, Moore SJ, Hawthorn JA, Neale MH, Terry LA. PrP(Sc) is associated with B cells in the blood of scrapie-infected sheep. Virology. 2010;**405**(1):110-119

[173] Mathiason CK, Hayes-Klug J, Hays SA, Powers J, Osborn DA, Dahmes SJ, et al. B cells and platelets harbor prion

infectivity in the blood of deer infected with chronic wasting disease. Journal of Virology. 2010;**84**(10):5097-5107

[174] Horiuchi M, Furuoka H, Kitamura N, Shinagaw M. Alymphoplasia mice are resistant to prion infection via oral route. The Japanese Journal of Veterinary Research. 2006;**53**(3-4):149-157

[175] van Keulen LJ, Bossers A, van Zijderveld F. TSE pathogenesis in cattle and sheep. Veterinary Research. 2008;**39**(4):24

[176] van Keulen LJ, Schreuder BE, Vromans ME, Langeveld JP, Smits MA. Pathogenesis of natural scrapie in sheep. Archives of Virology. Supplementum. 2000;(16):57-71

[177] van Keulen LJ, Vromans ME, van Zijderveld FG. Early and late pathogenesis of natural scrapie infection in sheep. APMIS. 2002;**110**(1):23-32

[178] Tabouret G, Lacroux C, Lugan S, Costes P, Corbiere F, Weisbecker JL, et al. Relevance of oral experimental challenge with classical scrapie in sheep. The Journal of General Virology. 2010;**91**(Pt 8):2139-2144

[179] Espenes A, Press C, Landsverk T, Tranulis M, Aleksandersen M, Gunnes G, et al. Detection of PrP(Sc) in Rectal Biopsy and Necropsy Samples from Sheep with Experimental Scrapie. Journal of Comparative Pathology. 2006;**134**(2-3):115-125

[180] Gonzalez L, Dagleish MP, Bellworthy SJ, Siso S, Stack MJ, Chaplin MJ, et al. Postmortem diagnosis of preclinical and clinical scrapie in sheep by the detection of disease-associated PrP in their rectal mucosa. The Veterinary Record. 2006;**158**(10):325-331

[181] Gonzalez L, Martin S, Siso S, Konold T, Ortiz-Pelaez A, Phelan L, et al. High prevalence of scrapie in a dairy goat herd: Tissue distribution of disease-associated PrP and effect of PRNP genotype and age. Veterinary Research. 2009;**40**(6):65

[182] Spraker TR, Gidlewski TL, Balachandran A, VerCauteren KC, Creekmore L, Munger RD. Detection of PrP(CWD) in postmortem rectal lymphoid tissues in Rocky Mountain elk (*Cervus elaphus* nelsoni) infected with chronic wasting disease. Journal of Veterinary Diagnostic Investigation. 2006;**18**(6):553-557

[183] Wolfe LL, Spraker TR, Gonzalez L, Dagleish MP, Sirochman TM, Brown JC, et al. PrPCWD in rectal lymphoid tissue of deer (Odocoileus spp.). The Journal of General Virology. 2007;**88** (Pt 7):2078-2082

[184] Thomsen BV, Schneider DA, O'Rourke KI, Gidlewski T, McLane J, Allen RW, et al. Diagnostic accuracy of rectal mucosa biopsy testing for chronic wasting disease within white-tailed deer (*Odocoileus virginianus*) herds in North America: Effects of age, sex, polymorphism at PRNP codon 96, and disease progression. Journal of Veterinary Diagnostic Investigation. 2012;**24**(5):878-887

[185] McCulloch L, Brown KL, Bradford BM, Hopkins J, Bailey M, Rajewsky K, et al. Follicular dendritic cell-specific prion protein (PrP) expression alone is sufficient to sustain prion infection in the spleen. PLoS Pathogens. 2011;7(12):e1002402

[186] Krautler NJ, Kana V, Kranich J, Tian Y, Perera D, Lemm D, et al. Follicular dendritic cells emerge from ubiquitous perivascular precursors. Cell. 2012;**150**(1):194-206

[187] Liu LM, MacPherson GG. Antigen acquisition by dendritic cells: Intestinal dendritic cells acquire antigen administered orally and

can prime naive T cells in vivo. The Journal of experimental medicine. 1993;**177**(5):1299-1307

[188] Banchereau J, Briere F, Caux C, Davoust J, Lebecque S, Liu YJ, et al. Immunobiology of dendritic cells. Annual Review of Immunology. 2000;**18**:767-811

[189] Cerovic V, Houston SA, Westlund J, Utriainen L, Davison ES, Scott CL, et al. Lymph-borne CD8alpha+ dendritic cells are uniquely able to cross-prime CD8+ T cells with antigen acquired from intestinal epithelial cells. Mucosal Immunology. 2015;**8**(1):38-48

[190] Wykes M, Pombo A, Jenkins C, MacPherson GG. Dendritic cells interact directly with naive B lymphocytes to transfer antigen and initiate class switching in a primary T-dependent response. Journal of Immunology. 1998;**161**(3):1313-1319

[191] Macpherson AJ, Uhr T. Induction of protective IgA by intestinal dendritic cells carrying commensal bacteria. Science. 2004;**303**(5664):1662-1665

[192] Saeki H, Wu MT, Olasz E, Hwang ST. A migratory population of skin-derived dendritic cells expresses CXCR5, responds to B lymphocyte chemoattractant in vitro, and co-localizes to B cell zones in lymph nodes in vivo. European Journal of Immunology. 2000;**30**(10):2808-2814

[193] Leon B, Ballesteros-Tato A, Browning JL, Dunn R, Randall TD, Lund FE. Regulation of T(H)2 development by CXCR5+ dendritic cells and lymphotoxin-expressing B cells. Nature Immunology. 2012;**13**(7):681-690

[194] Flores-Langarica A, Sebti Y, Mitchell D, Sim R, Macpherson G. Scrapie Pathogenesis: The Role of Complement C1q in Scrapie Agent Uptake by Conventional Dendritic Cells. The Journal of Immunology. 2009;**182**(3):1305-1313

[195] Gunn MD, Ngo VN, Ansel KM, Ekland EH, Cyster JG, Williams LT. A B-cell-homing chemokine made in lymphoid follicles activates Burkitt's lymphoma receptor-1. Nature. 1998;**391**(6669):799-803

[196] Ansel KM, Ngo VN, Hyman PL, Luther SA, Forster R, Sedgwick JD, et al. A chemokine-driven positive feedback loop organizes lymphoid follicles. Nature. 2000;**406**(6793):309-314

[197] Yu P, Wang Y, Chin RK, Martinez-Pomares L, Gordon S, Kosco-Vibois MH, et al. B cells control the migration of a subset of dendritic cells into B cell follicles via CXC chemokine ligand 13 in a lymphotoxin-dependent fashion. Journal of Immunology. 2002;**168**(10):5117-5123

[198] Bradford BM, Reizis B, Mabbott NA. Oral prion disease pathogenesis is impeded in the specific absence of CXCR5-expressing dendritic cells. Journal of Virology. 2017;**91**(10)

[199] Raymond CR, Aucouturier P, Mabbott NA. In vivo depletion of CD11c+ cells impairs scrapie agent neuroinvasion from the intestine. Journal of Immunology. 2007;**179**(11):7758-7766

[200] Sethi S, Kerksiek KM, Brocker T, Kretzschmar H. Role of the CD8+ dendritic cell subset in transmission of prions. Journal of Virology. 2007;**81**(9):4877-4880

[201] Cordier-Dirikoc S, Chabry J. Temporary depletion of CD11c+ dendritic cells delays lymphoinvasion after intraperitonal scrapie infection. Journal of Virology. 2008;**82**(17):8933-8936

[202] Wathne GJ, Kissenpfennig A, Malissen B, Zurzolo C, Mabbott NA.

Determining the role of mononuclear phagocytes in prion neuroinvasion from the skin. Journal of Leukocyte Biology. 2012;**91**(5):817-828

[203] Berney C, Herren S, Power CA, Gordon S, Martinez-Pomares L, Kosco-Vilbois MH. A member of the dendritic cell family that enters B cell follicles and stimulates primary antibody responses identified by a mannose receptor fusion protein. The Journal of Experimental Medicine. 1999;**190**(6):851-860

[204] Carp RI, Callahan SM. In vitro interaction of scrapie agent and mouse peritoneal macrophages. Intervirology. 1981;**16**(1):8-13

[205] Carp RI, Callahan SM. Effect of mouse peritoneal macrophages on scrapie infectivity during extended in vitro incubation. Intervirology. 1982;**17**(4):201-207

[206] Beringue V, Demoy M, Lasmezas CI, Gouritin B, Weingarten C, Deslys JP, et al. Role of spleen macrophages in the clearance of scrapie agent early in pathogenesis. The Journal of Pathology. 2000;**190**(4):495-502

[207] Maignien T, Shakweh M, Calvo P, Marce D, Sales N, Fattal E, et al. Role of gut macrophages in mice orally contaminated with scrapie or BSE. International Journal of Pharmaceutics. 2005;**298**(2):293-304

[208] Beekes M, Baldauf E, Diringer H. Sequential appearance and accumulation of pathognomonic markers in the central nervous system of hamsters orally infected with scrapie. The Journal of General Virology. 1996;**77**(Pt 8):1925-1934

[209] Beekes M, McBride PA, Baldauf E. Cerebral targeting indicates vagal spread of infection in hamsters fed with scrapie. The Journal of General Virology. 1998;**79**(Pt 3):601-607

[210] Baldauf E, Beekes M, Diringer H. Evidence for an alternative direct route of access for the scrapie agent to the brain bypassing the spinal cord. The Journal of General Virology. 1997;**78** (Pt 5):1187-1197

[211] McBride PA, Beekes M. Pathological PrP is abundant in sympathetic and sensory ganglia of hamsters fed with scrapie. Neuroscience Letters. 1999;**265**(2):135-138

[212] McBride PA, Schulz-Schaeffer WJ, Donaldson M, Bruce M, Diringer H, Kretzschmar HA, et al. Early spread of scrapie from the gastrointestinal tract to the central nervous system involves autonomic fibers of the splanchnic and vagus nerves. Journal of Virology. 2001;**75**(19):9320-9327

[213] Sowinski S, Jolly C, Berninghausen O, Purbhoo MA, Chauveau A, Kohler K, et al. Membrane nanotubes physically connect T cells over long distances presenting a novel route for HIV-1 transmission. Nature Cell Biology. 2008;**10**(2):211-219

[214] Aucouturier P, Geissmann F, Damotte D, Saborio GP, Meeker HC, Kascsak R, et al. Infected splenic dendritic cells are sufficient for prion transmission to the CNS in mouse scrapie. The Journal of clinical investigation. 2001;**108**(5):703-708

[215] Gousset K, Schiff E, Langevin C, Marijanovic Z, Caputo A, Browman DT, et al. Prions hijack tunnelling nanotubes for intercellular spread. Nature Cell Biology. 2009;**11**(3):328-336

[216] Langevin C, Gousset K, Costanzo M, Richard-Le Goff O, Zurzolo C. Characterization of the role of dendritic cells in prion transfer to primary neurons. The Biochemical Journal. 2010;**431**(2):189-198

[217] Tanaka Y, Sadaike T, Inoshima Y, Ishiguro N. Characterization of PrP(Sc)

transmission from immune cells to neuronal cells. Cellular Immunology. 2012;**279**(2):145-150

[218] Zhu S, Victoria GS, Marzo L, Ghosh R, Zurzolo C. Prion aggregates transfer through tunneling nanotubes in endocytic vesicles. Prion. 2015;**9**(2):125-135

[219] Fevrier B, Vilette D, Archer F, Loew D, Faigle W, Vidal M, et al. Cells release prions in association with exosomes. Proceedings of the National Academy of Sciences of the United States of America. 2004;**101**(26):9683-9688

[220] Castro-Seoane R, Hummerich H, Sweeting T, Tattum MH, Linehan JM, Fernandez de Marco M, et al. Plasmacytoid dendritic cells sequester high prion titres at early stages of prion infection. PLoS Pathogens. 2012;**8**(2):e1002538

[221] Jeffrey M, Halliday WG, Bell J, Johnston AR, MacLeod NK, Ingham C, et al. Synapse loss associated with abnormal PrP precedes neuronal degeneration in the scrapie-infected murine hippocampus. Neuropathology and Applied Neurobiology. 2000;**26**(1):41-54

[222] Jamieson E, Jeffrey M, Ironside JW, Fraser JR. Apoptosis and dendritic dysfunction precede prion protein accumulation in 87V scrapie. NeuroReport. 2001;**12**(10):2147-2153

[223] Cunningham C, Deacon R, Wells H, Boche D, Waters S, Diniz CP, et al. Synaptic changes characterize early behavioural signs in the ME7 model of murine prion disease. The European Journal of Neuroscience. 2003;**17**(10):2147-2155

[224] Siskova Z, Reynolds RA, O'Connor V, Perry VH. Brain region specific pre-synaptic and post-synaptic degeneration are early components of neuropathology in prion disease. PLoS One. 2013;**8**(1):e55004

[225] Betmouni S, Perry VH, Gordon JL. Evidence for an early inflammatory response in the central nervous system of mice with scrapie. Neuroscience. 1996;**74**(1):1-5

[226] Vincenti JE, Murphy L, Grabert K, McColl BW, Cancellotti E, Freeman TC, et al. Defining the microglia response during the time course of chronic neurodegeneration. Journal of Virology. 2015;**90**(6):3003-3017

[227] Falsig J, Julius C, Margalith I, Schwarz P, Heppner FL, Aguzzi A. A versatile prion replication assay in organotypic brain slices. Nature Neuroscience. 2008;**11**(1):109-117

[228] Zhu C, Herrmann US, Falsig J, Abakumova I, Nuvolone M, Schwarz P, et al. A neuroprotective role for microglia in prion diseases. The Journal of Experimental Medicine. 2016;**213**(6):1047-1059

[229] Minghetti L, Greco A, Cardone F, Puopolo M, Ladogana A, Almonti S, et al. Increased brain synthesis of prostaglandin E2 and F2-isoprostane in human and experimental transmissible spongiform encephalopathies. Journal of Neuropathology and Experimental Neurology. 2000;**59**(10):866-871

[230] Boche D, Cunningham C, Gauldie J, Perry VH. Transforming growth factor-beta 1-mediated neuroprotection against excitotoxic injury in vivo. Journal of Cerebral Blood Flow and Metabolism. 2003;**23**(10):1174-1182

[231] Thackray AM, McKenzie AN, Klein MA, Lauder A, Bujdoso R. Accelerated prion disease in the absence of interleukin-10. Journal of Virology. 2004;**78**(24):13697-13707

[232] Tamguney G, Giles K, Glidden DV, Lessard P, Wille H, Tremblay P,

et al. Genes contributing to prion pathogenesis. The Journal of General Virology. 2008;**89**(Pt 7):1777-1788

[233] Gomez-Nicola D, Fransen NL, Suzzi S, Perry VH. Regulation of microglial proliferation during chronic neurodegeneration. The Journal of Neuroscience. 2013;**33**(6):2481-2493

[234] Alibhai J, Blanco RA, Barria MA, Piccardo P, Caughey B, Perry VH, et al. Distribution of misfolded prion protein seeding activity alone does not predict regions of neurodegeneration. PLoS Biology. 2016;**14**(11):e1002579-e

[235] Gilch S. Overview of Chronic Wasting Disease. https://www.msdvetmanual.com/nervous-system/chronic-wasting-disease/overview-of-chronic-wasting-disease (accessed 25 April 2019)

[236] The Center for Food Security & Public Health. Chronic Wasting Disease. http://www.cfsph.iastate.edu/Factsheets/pdfs/chronic_wasting_disease.pdf (accessed 25 April 2019)

[237] Spraker TR, Miller MW, Williams ES, Getzy DM, Adrian WJ, Schoonveld GG, et al. Spongiform encephalopathy in free-ranging mule deer (*Odocoileus hemionus*), white-tailed deer (*Odocoileus virginianus*) and rocky mountain elk (*Cervus elaphus* nelsoni) in northcentral Colorado. Journal of Wildlife Diseases. 1997;**33**(1):1-6

[238] Vaccines OMoDTa, Animals fT. 2018. Available from: http://www.oie.int/standard-setting/terrestrial-manual/access-online/

[239] Authority EFS. The European Union summary report on surveillance for the presence of transmissible spongiform encephalopathies (TSEs) in 2017. EFSA Journal (published by John Wiley and Sons Ltd on behalf of European Food Safety Authority). 2018

[240] Andreoletti O, Budka H, Buncic S, Colin P, Collins JD, De Koeijer A, et al. Protocol for the evaluation of new rapid BSE post mortem tests. Opinion of the scientific panel on biological hazards. The European Food Safety Authority (EFSA) Journal. 2007;**508**:1-20

[241] Andreoletti O, Budka H, Buncic S, Colin P, Collins JD, De Koeijer A, et al. Protocol for the evaluation of rapid post mortem tests TSE in small ruminants. Opinion of the scientific panel on biological hazards. The European Food Safety Authority (EFSA) Journal. 2007;**509**:1-31

[242] Sigurdson CJ, Aguzzi A. Chronic wasting disease. Biochimica et Biophysica Acta. 2007;**1772**(6):610-618

[243] Xie Z, O'Rourke KI, Dong Z, Jenny AL, Langenberg JA, Belay ED, et al. Chronic wasting disease of elk and deer and Creutzfeldt-Jakob disease: Comparative analysis of the scrapie prion protein. The Journal of Biological Chemistry. 2006;**281**(7):4199-4206

[244] O'Rourke KI, Zhuang D, Lyda A, Gomez G, Williams ES, Tuo W, et al. Abundant PrP(CWD) in tonsil from mule deer with preclinical chronic wasting disease. Journal of Veterinary Diagnostic Investigation. 2003;**15**(4):320-323

[245] Andréoletti O, Orge L, Benestad SL, Beringue V, Litaise C, Simon S, et al. Atypical/Nor98 scrapie infectivity in sheep peripheral tissues. PLoS Pathogens. 2011;**7**(2):e1001285

[246] Jacobs JG, Langeveld JPM, Biacabe A-G, Acutis P-L, Polak MP, Gavier-Widen D, et al. Molecular discrimination of atypical bovine spongiform encephalopathy strains from a geographical region spanning a wide area in Europe. Journal of Clinical Microbiology. 2007;**45**(6):1821-1829

[247] (TSE) EEURLEf. TSE strain characterisation in small ruminants. A technical handbook for national reference laboratories in the EU. 2016. Available from: https://science.vla.gov.uk/tse-lab-net/documents/tse-oie-rl-handbook.pdf

[248] Daus ML, Breyer J, Wagenfuehr K, Wemheuer WM, Thomzig A, Schulz-Schaeffer WJ, et al. Presence and seeding activity of pathological prion protein (PrP(TSE)) in skeletal muscles of white-tailed deer infected with chronic wasting disease. PLoS One. 2011;**6**(4):e18345-e

[249] Pulford B, Spraker TR, Wyckoff AC, Meyerett C, Bender H, Ferguson A, et al. Detection of PrP_{CWD} In Feces From Naturally Exposed Rocky Mountain Elk (*Cervus elaphus nelsoni*) Using Protein Misfolding Cyclic Amplification. Journal of Wildlife Diseases. 2012;**48**:425-434

[250] Atarashi R, Wilham JM, Christensen L, Hughson AG, Moore RA, Johnson LM, et al. Simplified ultrasensitive prion detection by recombinant PrP conversion with shaking. Nature Methods. 2008;**5**:211

[251] Haley NJ, Richt JA. Evolution of diagnostic tests for chronic wasting disease, a naturally occurring prion disease of cervids. Pathogens (Basel, Switzerland). 2017;**6**(3):35

[252] Ricci A, Allende A, Bolton D, Chemaly M, Davies R, Fernández Escámez PS, et al. Chronic wasting disease (CWD) in cervids. EFSA Panel on Biological Hazards. 2017;**15**(1):e04667

[253] Samuel MD, Joly DO, Wild MA, Wright SD, Otis DL, Werge RW, et al. Surveillance Strategies for Detecting Chronic Wasting Disease in Free-ranging Deer and Elk: Results of a CWD Surveillance Workshop. Madison, WI: U.S. Geological Survey, National Wildlife Health Center; 2003. 43 p

[254] Uehlinger FD, Johnston AC, Bollinger TK, Waldner CL. Systematic review of management strategies to control chronic wasting disease in wild deer populations in North America. BMC Veterinary Research. 2016;**12**(1):173

[255] Miller MW, Williams ES, Hobbs NT, Wolfe LL. Environmental sources of prion transmission in mule deer. Emerging Infectious Diseases. 2004;**10**(6):1003-1006

[256] Miller MW, Fischer RJ. The First Five (or More) Decades of Chronic Wasting Disease: Lessons for the Five Decades to Come. Transactions of the North American Wildlife and Natural Resources Conference. 2016;**81**. http://cwd-info.org/wp-content/uploads/2018/12/81st-NAWNRC-Transactions_FINAL-CWD-Excerpt.pdf (accessed 25 April 2019)

[257] Manjerovic MB, Green ML, Mateus-Pinilla N, Novakofski J. The importance of localized culling in stabilizing chronic wasting disease prevalence in white-tailed deer populations. Preventive Veterinary Medicine. 2014;**113**(1):139-145

[258] Wolfe LL, Watry MK, Sirochman MA, Sirochman TM, Miller MW. Evaluation of a test and cull strategy for reducing prevalence of chronic wasting disease in mule deer (*Odocoileus hemionus*). Journal of Wildlife Diseases. 2018;**54**(3):511-519

Chapter 7

Infectious Disease Monitoring of European Bison (*Bison bonasus*)

Magdalena Larska and Michał K. Krzysiak

Abstract

In 2019, the 90th anniversary of the restitution of European bison (wisent) will be celebrated. Therefore, the chapter discusses the past, present, and future health threats of the *Bison bonasus* species that was on the edge of world extinction at the beginning of the twentieth century and was restituted with great efforts from many researchers, breeders, forestry workers, and caretakers. Due to the dramatic genetic "bottleneck" that depleted the gene pool, increasing the inbred of today's European bison, the breeding may face problems of decreased fertility, deficiency in growth, and increased susceptibility to diseases. While the increasing numbers of European bison may be enjoyed by breeders, the suitable habitat for the largest herbivore in Europe shrinks with increasing human population density, forestry, and agricultural activity. Additional threats include inappropriate management based on animal farming rather than sylvatic ecosystems, need for supplementary winter feeding, and establishment of breeding of related species such as American bison (*Bison bison*) in Europe. The control of European bison exposure to pathogens through passive and active surveillance is a key component of the species conservation. Hereby, the current knowledge on the epidemiology of the most significant infectious diseases in European bison is presented.

Keywords: wisent, surveillance, health, infection, pathogen

1. Introduction

The purpose of this work was to show the real and potential health risks of the European bison, which have or can influence the general condition of the population and affect the restitution effects of this endangered by extinction species. From the species history, through the old, however recently increasingly re-emerging, threats until the new challenges for the species conservation, we have tried to present the most complete picture of the veterinary aspect of the species protection. Not without significance is also the manner in which these tasks are implemented, i.e., active species protection, not always accepted, but effective, which is best evidenced by the continuous increase in the number of the world's European bison population. Today, when the number of the species exceeded 7000 individuals, it is not of key importance to protect individual European bison at all costs but to improve the welfare of the whole population, e.g., by minimizing the risk of infectious diseases by eliminating sick animals. Health threats can be removed or minimized through implemented prophylaxis and monitoring of pathological conditions in individual populations. Moreover, the exchange of experience and cooperation between breeders at national and international levels is extremely important. Another important aspect is the sanitary control of the animals in movement, especially of the greatest health threats such as tuberculosis and bluetongue disease and conducting quarantine for the introduced animals. One should also not underestimate the potential health hazards that can be assessed by postmortem diagnostics of fallen and selectively eliminated animals followed by discriminatory laboratory testing. The future of the European bison depends on the wise care of the herd managers and veterinary medicine specialists based on the best scientific and practical knowledge.

2. Species history and restoration

The European bison (*Bison bonasus*) is currently the wild land mammal of Europe, belonging to the cloven-hoofed ruminants, the family *Bovidae*. The males may reach up to 1000 kg (average 700–800 kg) of weight, while female up to 650 kg (average 400–500 kg) [1]. At the beginning of the last millennium, lowland European bison lived widely in the vast forests covering most of Europe and were highly valued hunting prize. In Europe, the wild bison survived the longest at the Białowieża Forest while it was within the borders of the Grand Duchy of Lithuania, the Kingdom of Poland, and the Polish-Lithuanian Commonwealth. In 1919, the last free-living European bison in the Białowieża Forest was killed. In 1927, the scientific expedition to Caucasus has also failed to find any living European bison there. The period between 1919 and 1929 was the time when not a single European bison was present at the Białowieża Forest. When in 1924 a decision was taken to try to save the species, only 54 European bison were still alive worldwide; however only a part of those animals participated in the species restitution. Currently, all European bison living in the world originate from 12 animals, referred to as the founding group. The European bison belongs to two lines, i.e., Lowland (derived only from seven individuals) and Lowland-Caucasian. As part of the restitution breeding in September 1929, the first European bison returned to Białowieża, but it was not until 1952 that two individuals were released into the wild [2]. Currently, over 1200 animals live in the entire Białowieża Primeval Forest (560 on the Belarusian side); there are almost 1900 of them all over Poland and over 7000 in the whole world [3]. In Poland, European bison appear in 5 free-living populations and over 20 enclosures (**Figure 1**). European bison are protected animals and are inscribed in the IUCN Red List of Threatened Species [4] under the vulnerable category. In Poland and Belarus, European bison have the status of animals under species protection, while in other countries they are treated as free-living animals kept in closed breeding. Since they need a space composed of a forest environment in which they spend most of their time, as well as grasslands (pastures) for feeding, only few places in Europe meet the conditions without causing losses in agricultural crops and forestry [1]. An efficient way to minimize conflicts is to carry out winter supplementary feeding of these animals with roughage (hay), which provides an existential minimum for European bison and may protect the crops [5]. For veterinarians, breeders, and carers of the free-living bison population, winter gathering is the appropriate occasion to overview the herd. All European bison are counted, and the sexual (male, female) and age (calves, 0–1 year; youth, 2–3 years; adult, 4 years and more) structures of the herd are determined. The individuals are entered into the European Bison Pedigree Book [3]. In addition, the winter overview of the herd allows to identify any diseased or weak animals, which may constitute a potential epizootic threat to other European bison [1, 6]. Because European bison remain endangered and rare species, as well as attractive because of their uniqueness as a show species, breeding restitution of this species is also carried out in enclosures and reserves in Poland and Europe. All European bison kept in captivity have a

DOI: http://dx.doi.org/10.5772/intechopen.84290

Figure 1.
European bison population distribution in Poland, 2018 (graphic design by J. Tomana).

unique pedigree number and a name, specific for the country and place of the origin [3]. Keeping breeding centers allows to control the breeding of European bison by selection of animals with known pedigree and least related. Moreover, breeding enclosures provide a genetic and breeding reserve of pedigree animals in case if any depopulating incidence occurs, what the species had already experienced at the beginning of twentieth century. Therefore, the management of the enclosures needs to be supervised by a veterinarian. Each movement of animals between breeding centers should be preceded by a clinical examination of the animal; a laboratory test for the most important infectious diseases and prophylaxis (more often therapeutic in practice) against ecto- and endoparasites should be applied [7].

One of the dangers, often underestimated by ignorance or financial incentives, is the commercial breeding of American bison (*Bison bison*) in Europe, which is a serious genetic and epizootic threat for European species. The European and American bison crossbreed can naturally reproduce, which means that the co-existence of both species in one area could lead to loss of the protected species genetic purity and threaten biodiversity and conservation [8]. Therefore, the establishment of commercial American bison farms disqualifies the immediate vicinity as potential places for the release of European bison into the wild. Moreover, the presence of American bison poses a critical epizootic risk to the European cousin, since they can transmit the pathogens threating European bison health, which are absent or emerging in Europe [9], for example, *Fasciola magna* invasions [10] or epizootic hemorrhagic disease, caused by EHDV [11]. The close interspecies relationship does not ensure similar susceptibility to diseases. Significant differences in clinical outcome of foot-and-mouth disease virus (FMDV) infection were demonstrated, for instance, between dromedary and Bactrian camels [12]. The burden of FMD epizootics in Polish European bison at the beginning of the twentieth century was much higher than in cattle in the same area [13, 14].

3. European bison health threats of the twentieth century

Despite that the species was just starting to be re-established, European bison were put under epidemic pressure and exposed to many pathogens endemic to domestic ruminants in Poland at the beginning of the twentieth century. After the Second World War, the European bison population in Poland systematically increased, except for several incidents associated with the occurrence of infectious diseases (**Figure 2**). Konrad Wróblewski [14], a vet and researcher, gave the first detailed descriptions of the health problems in European bison in Białowieża at the turn of the nineteenth and twentieth centuries. At that time, the causes of morbidity and pathological changes were not yet fully understood due to the lack of knowledge and research tools. Dr. Wróblewski observed pleuropneumonia caused by *Pasteurella multocida* and connected to the *Strongylidae* invasions of lungs [15]. The problem of purulent pneumonia is still observed in European bison at Białowieża [6]; however the agent, *Pasteurella multocida subsp. multocida* serogroup A was identified just recently [16]. At the time, the highest mortality in European bison was caused by blackleg. The infection with *Clostridium chauvoei* probably caused death of 172 European bison in 1904 [14]. The most common were *Fasciola hepatica* and *Trypanosoma* parasite invasions, which affected health condition causing weakness, weight loss, and secondary infections. Occurrence of Tryp*anosoma wrublewski* protozoans in the blood of European bison was reported still in the 1990s [15]. Another problem was FMD, endemic in cattle from the area; nevertheless Dr. Wróblewski described only 5% mortality in European bison in the 1920s. However, FMD was taking its toll in the 1950s at the south of Poland causing almost complete depopulation of European bison reserves in Pszczyna, Niepołomice, and Gorce [17–19]. Staśkiewicz [13] has observed FMD in American bison and European-American bison hybrids bred in Smardzewice reserve in 1938. He has noticed significantly higher resistance to FMDV infection of those animals in comparison to European species. This also contends against the introduction of this invasive to Europe bison species into the areas where European bison are being reared.

In 1980s, due to the occurrence of Q fever in domestic animals in northeastern Poland, several dozen free-ranging European bison from Borecka Forest were

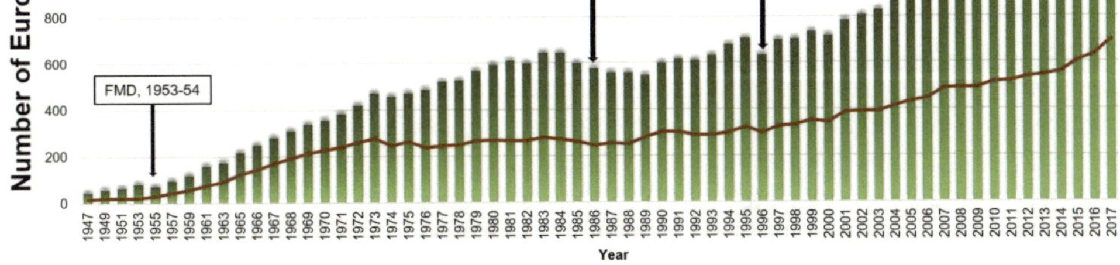

Figure 2.
Dynamics of the European bison number in Poland (bars) and Białowieża Forest (curve) between 1947 and 2017 (European Bison Pedigree Book). The cases of diseases which caused significant declines in the population size and dates of their detection are marked.

DOI: http://dx.doi.org/10.5772/intechopen.84290

examined for the presence of specific antibodies [20]. The high seroprevalence (76%) to *Coxiella burnetii* found in European bison indicated that they could be a potential reservoir of Q fever for humans and other animals. The European bison-human transmission was suspected, since the infection was also confirmed in some employees (some hospitalized) of the Białowieża National Park (BNP) staff. Szarek et al. [21] have linked *C. burnetii* infection with the pathomorphological changes of the heart and kidneys observed in European bison specific for Q fever.

The first signs of another important disease in the male European bison, referred to as necrotic posthitis or balanoposthitis (called also pizzle rot), the causes of which still remain unsolved, were observed in Białowieża in 1980 remain unsolved, were observed in Białowieża in 1980 (**Figure 3**) [22]. *Trueperella pyogenes* and some other purulent bacteria have been isolated from the diphtheroid-necrotic changes observed on the prepuce and penis, which may lead to self-mutation of the penis in severe cases causing primarily great pain and immunosuppression, excluding the affected males from breeding [6, 22–24]. Infection with BoHV-1 causing similar symptoms in cattle was excluded from the investigation, since hardly any European bison affected had detectable antibodies to the virus and no isolation from the tissues was possible [1, 22, 25]. Also no association between *Trypanosoma* invasion and balanoposthitis was found [1]. Prior to the prepuce and penis lesions reported in Białowieża, similar symptoms were observed in Pszczyna, in the eastern part of Białowieża Forest (now Belarus), Ukraine, and Russia [1]. The epidemic of balanoposthitis in Białowieża included annually up to 15.2% males in 1993, with the average of 6.1% between 1980 and 2015. Selective culling was found the most effective method to decrease the cases, especially in the recent years [1, 6]. The symptoms are observed in all age groups of males, including calves and 2–3-year-old European bison males, yet not reproductively active [1, 6, 23]; therefore venereal route of transmission may rather be rejected. Since most cases occurred at Białowieża Forest, involvement of some environmental or vectors such as ticks was speculated, however, never verified. Since European bison genetic pool had gone through a dramatic "bottleneck" at the beginning of the twentieth century, high inbred was suspected to play a role; however, little evidence was reported [26, 27].

One of the most devastating bacterial diseases, which remains a current problem in Polish European bison populations, is tuberculosis. The disease was diagnosed in the free-living European bison population in the Bieszczady mountains in the 1990s [28–31], European bison are very susceptible to mycobacterial

Figure 3.
Male Bison bonasus with clinical signs of balanoposthitis showing symptoms of impeded urination (stranguria) and swelling of the whole external genital area (photo: Ł. J. Mazurek).

infections, and since no vaccination programs or treatment could have been introduced, the only method of disease control and prevention from spreading was the elimination of infected European bison [28, 29, 31–34]. The eradication resulted in a significant depopulation observed in decrease of the numbers of European bison (**Figure 2**). Tuberculosis affecting European bison in Poland is caused by *Mycobacterium caprae* [35]. The management of the diseases is extremely difficult especially in wildlife, since the mycobacteria transmit readily between different wild species such wild boar, deer, and wolves in Bieszczady mountains [31]. American bison bred in Poland are also considered important reservoir of tuberculosis [36].

4. Telemetric collars and European bison movement monitoring

Free-ranging European bison create herds consisting of cows, their offspring, and youth; solitary bulls live outside of the herd and, however, approach or join in the mixed groups during rutting season and winter (for feeding), while young bulls often create small male groups [37]. Since European bison move over considerable distances during the day (especially the bulls), in order to monitor the distribution of these animals and study their use of space and different ecosystems, telemetric transmitters are being placed on selected individuals. Through such transmitters, we can track the daily rhythm of European bison and the use of various habitats for feeding and resting, divided into seasons and various periods relevant to the behavior of animals (the calving season, mating season) [1]. Such transmitters can use a radio signal emission (the oldest way, currently replaced by GPS); it requires the involvement of a person with a radio receiver together with an antenna to track the animal and record their location. Another way is to connect the transmitter to the server via cellular telephony; coordinates from the GPS system are collected, converted, and transferred to the operator's server, processed and plotted on the map; however, it requires the animal remains within the GSM network, and the reading may be analyzed by the inspector only after some time. The most modern, but also significantly more expensive, way is placing satellite emitters on the European bison, which allows tracking these animals in different conditions (also outside the GSM network) and the ability to record the location of the "marked" animal systematically regardless of the environment in which it is located. By monitoring European bison for a longer period of time, maps can be created for the use of both males and females of different ecosystems, determine the range of existence of individual herds, and determine, for example, the nutritional preferences of these animals. The telemetric coordinates of European bison location are applied to maps and visualized showing land usage of individuals/herds in relation to season, differences in distribution, and behavior of females and males [38]. Same as European bison counting, setting up telemetry transmitters is usually performed during winter concentration at feeding places. For this purpose, the individual is pharmacologically immobilized (**Figure 4**) [7], and the belt with the telemetry transmitter is adjusted to its neck. This is important because a too tight collar may cause difficulties in swallowing and regurgitation, while too loosely attached collar may cause injuries (entrapment of the limb, foreign bodies such as branches). For an adult European bison, a telemetry transmitter that weighs approx. 2 kg may be compared to a watch on a human wrist. Additionally, during the pharmacological immobilization of European bison, samples are collected for veterinary, genetic, and toxicological tests, which is a part of the health monitoring described in paragraph 3.

Figure 4.
Chemical immobilization of free-living European bison (photo: J. Klej).

5. Current major epidemic threats and monitoring of European bison

When considering the problems of infectious and invasive diseases, three aspects (the so-called epidemiological triangle: animal-environment-pathogen) should be considered. The epizootic and epidemiological situation is a resultant of the interactions between the natural environment of the free-living animals, other wild reservoir species and farm animals at the wildlife-lifestock interface. An important environmental component will also be access to competent vectors of a given microorganism. In the case of environmental impact, it is also necessary to take into account people who are susceptible to zoonoses, can themselves be a source of infections, or can be a mechanical vector of infectious and invasive agents transmitted to animals, becoming endemic in sylvatic environment [39]. Those aspects meet in the One Health approach, which include wildlife as a key component of the ecosystem [9, 32, 40, 41]. Therefore, in the case of an epizootic, surveillance should be carried out in both domestic and non-domestic populations, both free-living and captive, and the potential role of humans (animal care takers, breeders, vets) as vectors should be taken into account. Climate, environment, socioeconomical changes create alterations also in the distribution of infectious diseases and remind us that animals and humans, environment, and pathogens are the elements of the same ecosystem.

In the last 20 years, the population size of European bison in Białowieża Forest and in the whole country increased more than twice (**Figure 2**). It generated an increasing epidemiological pressure connected to higher animal density, a need for expanding the habitat, and what further increases the frequency of contacts between wildlife, farm animals, and humans. The evident environmental changes may provoke also emergence and re-emergence of new pathogens, change the seasonality of wild species, and induce stress and therefore the immunosuppression leading to increased susceptibility to diseases.

Since the European bison population was on the verge of extinction at the beginning of the twentieth century, each individual was extremely valuable; therefore one of the main aspects of European bison restitution still remains veterinary health protection, especially in terms of the threat of infectious and invasive disease occurrence. The role of European bison veterinarians and keepers include the supervision, health monitoring, understanding disease characteristics, identification of disease risks, provision of information for control, prevention or treatment, and evaluation of the effectiveness of control and its adjustment. One of the effective tools to control the disease is selective elimination of diseased animals and limiting the population size to reduce its density [32]. The selection is made by a panel of experts in the field; however, it should be remembered that the approach presented here is based on veterinary health protection sometimes argued by epidemiologists [42]. In the case of protected species such European bison, enforcement of the veterinary regulations to protect the species by a sacrifice of few diseased individuals, which are reservoirs of the pathogens putting on risk the whole population, becomes often inconvenient decision against some environmentalist protests.

European bison monitoring in Poland consists of four elements: (1) passive surveillance, (2) active (targeted) monitoring, (3) sanitary control of transported animals, and (4) habitat monitoring (**Figure 5**) [43]. The first approach is based on one of the most suitable ways of health control, the postmortem examination as an element of mortality and morbidity investigations and sampling of the material for laboratory testing [6, 23, 44].

The most recent study involved the analysis of necropsy findings of over 230 fallen or selectively culled European bison from Białowieża Forest between 2008 and 2013 [6]. Changes in the male reproductive tract such as posthitis and balanoposthitis remained the most common pathological changes observed in European bison bulls; however, it should be kept in mind that those changes are the main reason for bull elimination (**Figure 6**). Moreover, the decrease in the proportion of males with prepuce and penis lesions of the whole Białowieża male population in regard to the last century was significant [1, 44].

The next most common problem included pneumonia (45% necropsied animals) and pulmonary emphysema (33%) [6], which were also reported in earlier studies [44]. Respiratory problems were associated with *Dictyocaulus viviparus* invasion of the lungs observed macroscopically; however, the characteristic picture of interstitial or catarrhal-purulent pneumonia suggested also involvement of some infectious agents, what was confirmed for *Pasteurella multocida* [16]. The study has shown that the liver fluke *Fasciola hepatica* still remained endemic in European bison in Białowieża [6]. The presence of the parasite is associated directly with the habitat of European bison in Białowieża [45]. The most frequent cause of death of European bison under the age of 6 months were body injuries caused by other European bison or less often as a result of other accidents. An important finding was that the frequency of pathological changes was higher in the selectively culled European bison than those that were found dead in the wild. This confirms the purposefulness of selection and elimination of diseased individuals as a tool to improve the health and welfare of the population of the species and allows to limit the number

DOI: http://dx.doi.org/10.5772/intechopen.84290

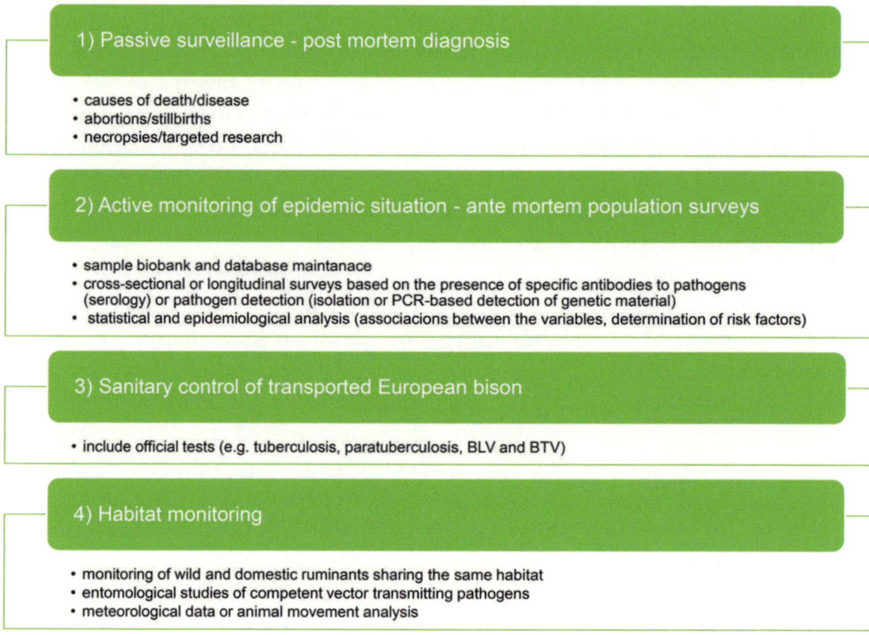

Figure 5.
The four main elements of the European bison health monitoring scheme in Poland.

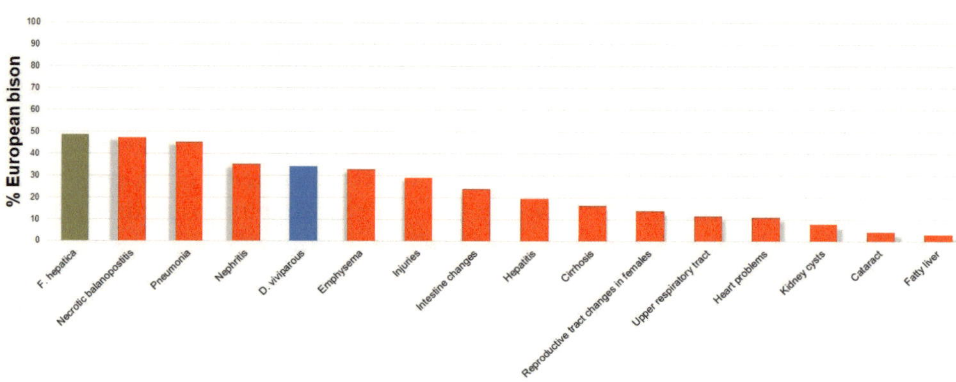

Figure 6.
Percentage of European bison with individual pathological changes and parasitic invasions observed at postmortem examinations performed between 2008 and 2013 in Białowieża Forest [6].

of animals that constitute a potential reservoir of infectious and invasive diseases [32]. Additionally, unlike farm animals, European bison are protected animals and live to a mature age; the pathological changes characteristic of elderly individuals, such as kidney cysts, fatty liver, and cataracts may be observed [6]. However, their frequency in the studied European bison did not exceed several percent.

Another large serosurvey has revealed the possible involvement of some respiratory viruses in the pathogenesis of the frequent changes observed in the lungs or in the upper respiratory tract such as bovine adenovirus type 3 (BAdV-3), bovine parainfluenza type 3 (PIV-3), and bovine respiratory syncytial virus (BRSV) [46]. The high seroprevalences observed especially for BAdV-3 (60%) and PIV-3 (34%) were surprising, also because they have not been studied too intensively before [47]. Interestingly, the association between BAdV-3 and PIV-3 infections and health status of European bison were demonstrated, with significantly lowest seroprevalences in the apparently healthy animals. Whether the high BAdV-3 and PIV-3 infection rates were due to the circulation inside the European bison population or it is connected to the transmission of those viruses from domestic ruminants remains unsolved. However, higher infection rates in the free-living European bison suggested that it might have been associated with a spillover from farm animals. The European bison kept in the enclosures were possibly protected from a direct contact with domestic species, not sharing grazing areas as observed in Białowieża [48, 49]. The studies on the characterization of the viral strains in European bison, what may explain the transmission source, are ongoing.

Moreover, other endemic in ruminants infectious agents were considered as potential threats to European bison. Those included pestiviruses with bovine viral diarrhea virus type 1 (BVDV-1) and alpha- and gammaherpesviruses (bovine herpesvirus type 1—BoHV-1, BoHV-4, and BoHV-6), endemic in Polish cattle herds [46]. While in farm animals those viruses are responsible for economic losses, they may prevent the reproduction of an endangered species and make the restitution program fail. However, very low seroprevalence to those viruses suggests certain resistance or only accidental exposure of European bison and hence little importance for European bison at present [15, 25, 46, 47]. On the other hand, low seroprevalence means that most European bison are naïve and fully susceptible to infections with BVDV-1 and BoHV-1, BoHV-4, and BoHV-6. The involvement of herpesviruses in the etiology of balanoposthitis was also disclosed [22].

Other respiratory agents, which could affect the clinical picture observed frequently postmortem in European bison such as *Mycoplasma* spp. and *Mycobacterium* spp., are probably of limited geographical importance [33, 46, 50, 51]. Infections with *Mycoplasma bovis* in few individuals suggest that the exposure was accidental and, however, might have added to some pathological changes of the respiratory tract observed in one case such as intrabronchial pneumonia and pleurisy [51]. In addition, *Mycoplasma bovigenitalium* was suggested to be involved in balanoposthitis under serological evidence; however, the bacteria have never been detected in the lesions of the male genital tract [50]. Moreover, the occurrence of *Mycoplasma ovipneumoniae* infections was never investigated; notwithstanding this direction is worth considering in the pathogenesis of respiratory disorders [52]. Bovine tuberculosis (bTB) remains one of the major infections putting on risk the whole population. *Mycobacterium* accompanies European bison since the dawn of time. Bacterial DNA has been discovered in the bison fossil dated 17,000 years before the present [53]. One of the major obstacles for tuberculosis control in wildlife is that the mycobacteria spread among different mammalian species and survive in numerous biotopes (soil, water, vegetation) becoming endemic in the wild ecosystem for a very long time. Additionally, the diagnostic test used for European bison is dedicated mostly to cattle; therefore, the sensitivity may vary. Sample quality is critical, and the use of several complementary diagnostic methods is necessary to confirm a bTB case in European bison [54]. Fortunately, the disease has not been observed in European bison from the world largest population in Białowieża for some decades now, probably owing to the elimination of the pathogen from cattle in Poland. The disease stroke, however, in a reserve in the European bison breeding center in Smardzewice [33, 46], what has led to stamping out of the whole herd in 2018. The outbreak as the previous one in Bieszczady mountains was connected to *Mycobacterium caprae* [31, 35]. Reports of tuberculosis in captive European bison caused by *M. bovis* in German and Brazilian zoos underline the need for control in the species being important bTB reservoir worldwide [55–57].

Further studies are concerned with pathogens, which may be depopulating through their influence on European bison reproduction, reducing already fragile number of the species under restitution. Low seroprevalences of *Brucella* spp., *Leptospira* spp., and *Toxoplasma gondii* observed in European bison in eight main Polish populations sampled between 2011 and 2015 by Krzysiak et al. [46] suggest limited importance of those agents for the species conservation at present. It had been also confirmed by the results of some previous investigations [47, 58]. Only slightly higher *T. gondii* parasite seroprevalence was observed in European bison from Białowieża prior to the study of Krzysiak et al. [46, 59]. It was followed by the report of *T. gondii* isolation from an aborted at 4–5-month-old gestation of European bison fetus in Białowieża in 2014 [60]. Similar to *T. gondii*, European bison pose some potential as *Neospora caninum* and *Chlamydia* spp. reservoirs [58, 61].

Bovine leukemia virus (BLV) and *Mycobacterium avium subsp. paratuberculosis* (Johne's disease) infections, endemic in Polish cattle, are subject of official tests performed routinely in the movement of European bison inside the European Union. Only one case of subclinical BLV infection has been reported so far in the 1880s. Just as paratuberculosis [51], bovine leukemia does not pose a serious threat to Polish European bison at present.

Finally, it is worth mentioning European bison susceptibility to malignant catarrhal fever virus (MCFV) infection, which is comparable to cattle's. The reports of clinical picture of MCF in the species kept in zoos are quite old [62, 63]. MCF is considered more emerging and non-endemic in Poland; therefore, no current data exists. However, it was well documented that the exposure of American bison to sheep-associated MCFV is frequently fatal [64, 65]. In 2018, an outbreak of fatal MCF in European bison enclosure in Switzerland was consulted with European Bison Pedigree Book department at the Białowieża National Park; however no report was published officially (Baumgartner, personal communication).

6. Emerging, re-emerging, and vector-borne pathogens infecting European bison

The twenty-first century brings new challenges of the protection of animal health, including free living. Social and economic changes, globalization, intensification of intra- and intercontinental trade and travels, and environment and climate change contributed to the observed increase in the risk of emergence and re-emergence of pathogens [66]. The last few years have brought an increase in the importance of new pathogens, completely unknown before or not present in a given geographical latitude [67, 68], which may also have a significant impact on the health of an endangered species such as the European bison. It is related to processes related to human activity or changes in the environment but also to the development of science, improvement of disease diagnostics, and the introduction of new cognitive techniques, such as metagenomics and next-generation sequencing (NGS). Due to the habitat and the maintenance of a part of the European bison population in captivity, they are exposed to pathogens from both wild and domestic animals. Along with climate change, infections caused by pathogens transmitted by vectors such as arthropods (ticks, biting midges, mosquitoes) are more and more frequently reported in our part of Europe. In 2007, the first cases of infection with bluetongue virus (BTV) serotype 8 (BTV-8) causing significant mortality in European bison at an enclosure in Hardehausen occurred. The virus is transmitted by the blood-sucking midges of the family of *Culicoides* spp. (**Figure 7a, b**). BTV infections have never been detected as far north of Europe. During the epizootics more than 30% of European bison kept in Hardehausen died, and the surviving animals received a vaccine containing homologous inactivated BTV-8 [69]. BTV-8 epizootics did not reach Poland; however another strain, BTV-14, was detected almost simultaneously in cattle and in European bison at the northeast of Poland, near the Belarusian and Lithuanian borders [11, 70].

The source remains puzzling, particularily since the virus was highly homologous to South African BTV-14 reference strains used in vaccines, which were never used in Europe. However, closely related BTV-14 was detected in Russia almost at the same time. Therefore, combining this with the dynamics of geographical spread of BTV-14 in Poland, use of some illegal vaccine containing or contaminated with this BTV strain was suggested [71]. Serosurveillance of BTV in European bison showed that the infections occurred only in the eastern populations [11]. Interestingly, despite that the discovered BTV-14 has been described as attenuated strain causing only subclinical infection in cattle, significant association between health status of European bison and the presence of specific antibodies was observed [11]. The exposure rate to BTV in selectively eliminated and fallen animals in respect to healthy ones was clear; however multivariable analysis has not confirmed the health status as a risk factor. Notwithstanding, the most important consequences of the occurring of BT are the culling of diseases or suspected animals and restrictions on the movement of animals, since the disease is notifiable disease listed by the World Organization. Control or eradication of BTV in protected wild species such as European bison is rather doomed, especially that the environment provides more susceptible species such as cervids and favorable conditions for midge reproduction and survival [72, 73]. BTV epidemiological situation remains under continuous supervision (unpublished data). Additionally, the occurrence of epizootic hemorrhagic disease virus (EHDV) infections transmitted by *Culicoides* spp. in European bison in Poland was also ruled out [11]. However, since some diseases of wildlife considered so far as absent in certain geographic areas such as African swine fever in wild boar or chronic wasting disease (CWD) in cervids become endemic in Europe, the threat of EHD in the future should also be considered real [9].

After the bluetongue epizootics (BTV) in 2007 [67], a new pathogen named Schmallenberg virus (SBV) was identified in Europe in 2011 [74], which spread very quickly across the continent. The first SBV infection in cattle in Poland was found in 2012, and in the same year, virus transmission to the ruminants of free-living ruminants in the Białowieża Forest, including European bison, was observed [75]. The first case of acute SBV infection in wild animals was actually confirmed in a fallen elk calf rescued by the animal keepers of the Białowieża National Park [76, 77]. This was the first report on the identification of the SBV virus in a free-living animal that began an international discussion on the importance of free-living animals as a reservoir of this new virus [77]. It proceeded the subsequent surveillance of European bison and other

Figure 7.
BTV, EHDV, and SBV insect vectors: biting midges of Culicoides obsoletus species—(a) freshly blood-fed and (b) gravid (containing a mature egg batch in the abdomen). Ultraviolet (UV) light trap (CDC 1212, John W. Hock Company, USA) for Culicoides spp. midges collection set at the Show Reserve in Białowieża National Park (c) (photo: Ł. Mazurek). The midges are active at night; therefore the trap was set on from dusk till dawn once a week during the vector activity season (April-October). UV light attracts the insects, which were trapped and fell down into the jar containing water with detergent [73].

wild ruminants at Białowieża Forest, what allowed to determined that the transmission of SBV occurred during 2012 *Culicoides* activity season [76]. Further studies concentrated on the epidemiology of SBV in European bison and the environment (other susceptible species, vector) [11, 73, 78]. SBV seroprevalence in European bison in all studies was significantly much higher than observed in the cervids sampled simultaneously at the same locations. It is not, however, due to the higher susceptibility of the European bison to SBV infection, rather than higher exposure to midge biting. Similar differences were reported between cattle and small ruminants. Most probably, higher exposure in larger ruminants is associated with their higher production of carbon dioxide, one of the strongest attractants of many arthropods including *Culicoides* spp. [79, 80]. SBV *in utero* infections may lead to congenital malformations of newborns, stillbirths, or abortions. Limited access to the material of aborted fetus in the wild European bison did not allow to draw many conclusions [73]; however some losses in reproduction rate may be expected as observed in cattle.

Since the transmission of arthropod-borne pathogens involves environmental factors, entomological and virological studies of the midges near European bison were performed between 2014 and 2015 [73]. Using specially designed ultraviolet (UV) light traps (**Figure 7c**) placed near European bison resting places at Białowieża National Park abundance, species composition and virus exposure were tested. The wood midges of *Culicoides achrayi* species were the most abundant with almost 280 thousand (over 50% of the whole midge number collected) individual midges of the species trapped over the 2-year period [73]. The presence of SBV was confirmed in some midges trapped in 2015. The work except its originality provided very needed data to study all *Culicoides*-borne infections including BTV in the sylvatic habitat of European bison.

Further studies suggest that European bison may be an important reservoir of tick-borne Lyme disease, since *Borrelia burgdorferi* was detected in blood of several animals in Białowieża [81]. Earlier, the presence of specific antibodies to *B. burgdorferi* was confirmed in European bison in the same area in the 1990s [82]. Other documented in European bison investigations of the infections transmitted by ticks include tick-borne encephalitis (TBE), tularemia, and anaplasmosis. While TBE virus was detected in the ticks collected from European bison [83], no antibodies supported the hypothesis that the species is an important reservoir of the virus [84]. European bison probably have little importance in the transmission of another zoonotic pathogen, *Francisella tularensis*, since none of the 251 individuals from eight Polish populations had antibodies against the bacteria according to Krzysiak et al. [46]. However, it is different when *Anaplasma phagocytophilum* responsible for human granulocytic ehrlichiosis (HGE) is considered. This tick-transmitted bacterium was found in a blood of over 66% of European bison in Białowieża Forest [85]. The pathogen was also confirmed to be present in the ticks collected from the animals [86]. However, whether any of the detailed tick-borne pathogens influences European health remains unexplored.

The last emerging disease with high epizootic potential we would like to discuss shortly is hepatitis E. The pigs including wild boar are the main reservoirs of hepatitis E virus (HEV); however, cases of virus to humans through cervid meat have been also reported. In a recent study, none of the European bison had antibodies against HEV [87]; however, the sensitivity of the methods to ruminants is debatable, and therefore the studies are being continued.

7. Conclusions

European bison survived to the present times, only thanks to human care and protection. Because Europe, unlike the Americas, was significantly more populated in the Middle Ages, European bison conservation by the Polish kings saved this species from total extinction, as was the case outside the Kingdom of Poland. Successive rulers continued to protect the species, but warfare led to the total depopulation of the last surviving free-living population in the Białowieża Primeval Forest. Thanks to the efforts of scientists and naturalists, the current global population of European bison which was founded by 12 individuals is over 7000. This is undoubtedly a success, but without the supervision of the health of European bison, it would be impossible. European bison, in general opinion, is characterized by a decreased resistance to infections due to their low genetic heterogeneity after passing through the "bottleneck." Except for *Mycobacterium* spp. and foot-and-mouth disease, there are few threats threatening to total depopulation of the species at the moment. However, many European bison seems to be susceptible to many pathogens which may influence their health status and exclude genetically valuable individuals from the breeding. Monitoring of the exposure is essential for the knowledge but also for taking the necessary steps to protect and control. It requires a sampling plan, which allows collection of representative number of specimen necessary for statistical analysis, which is not always feasible since European bison are few and most live in the wild. Therefore, many discussed above studies do not meet any epidemiological assumption and need to be analyzed with caution. Many retrospective studies suggest that European bison may be a reservoir of pathogens; however, the susceptibility to the infection is arguable, since they do not manifest any clinical symptoms (borreliosis, toxoplasmosis). In contrast to high-production animals, it seems not to be highly susceptible to infection, e.g., with pestiviruses and herpesviruses. At the same time, the possible interspecies transmission between domestic ruminants and European bison may be possibly responsible for the high prevalence of respiratory viruses and *Pasteurella multocida* infections in which clinical manifestations are observed frequently postmortally in European bison. Changing environment brings new challenges to European bison protection too, since the entire population may be at risk due to the emerging and re-emerging diseases observed recently such as bluetongue disease and malignant catarrhal fever. Therefore, maintaining the health monitoring of free-living European bison and eliminating all epizootic threats from the environment are crucial. It is also very important to follow the veterinary procedures and appropriate prophylactic measures when moving these animals. In our opinion, "prevent to protect" should stand for active protection, which proved to be the most effective in European bison restitution so far. Restitution breeding is a long and tedious process, often associated with making difficult and unpopular decisions.

Acknowledgements

The publishing of the chapter was funded by KNOW (Leading National Research Centre) Scientific Consortium "Healthy Animal-Safe Food," decision of Ministry of Science and Higher Education no. 05-1/KNOW2/2015. The collection of some data was funded from the project: '*In situ* conservation of the European bison in Poland - the north-eastern part' cofinanced by Operational Program Infrastructure and Environment (contract no: POIS.05.01.00-00-229/09-00).

Conflict of interest

We declare no conflict of interest exists for this publication.

DOI: http://dx.doi.org/10.5772/intechopen.84290

Author details

Magdalena Larska[1*] and Michał K. Krzysiak[2,3]

1 Department of Virology, National Veterinary Research Institute, Puławy, Poland

2 Białowieża National Park, Białowieża, Poland

3 Department of Food Hygiene and Public Health Protection, Faculty of Veterinary Medicine, Warsaw University of Life Sciences, Warszawa, Poland

*Address all correspondence to: m.larska@piwet.pulawy.pl

References

[1] Krasińska M, Krasiński ZŻ. Monografia Przyrodnicza. Białowieża: Chyra.pl; 2017. 448 p

[2] Krzysiak MK, Larska M, Dackiewicz J, Anusz K. Restitution of European bison (*Bison bonasus*) in Białowieża and the most common health problems of the species in the wild and captive populations in Poland. Parki Narodowe i Rezerwaty Przyrody. 2018;**37**:85-91

[3] Raczyński J, Bołbot M. European Bison Pedigree Book. Białowieża: Białowieski Park Narodowy; 2018. 84 p

[4] IUCN. 2008. IUCN Red List of Threatened Species. Available from: http://www.iucnredlist.org [Accessed: Dec 21, 2018]

[5] Olech W, Bielecki W, Bołbot A, Bukowczyk I, Dackiewicz J, Dymnicka M, et al. Hodowla żubrów. Poradnik utrzymania w niewoli. Warszawa: Stowarzyszenie Miłośników Żubrów; 2008. 100 p

[6] Krzysiak MK, Dackiewicz J, Kęsik-Maliszewska J, Larska M. Post-mortem evaluation of pathological lesions in European bison (*Bison bonasus*) in the Białowieża Primeval Forest between 2008 and 2013. Bulletin of the Veterinary Institute in Pulawy. 2014;**58**:421-431. DOI: 10.2478/bvip-2014-0065

[7] Krzysiak MK, Larska M. Pharmacological immobilization of European bison (*Bison bonasus*). Medycyna Weterynaryjna. 2014;**70**:172-175

[8] Olech W. Wpływ inbredu osobniczego i inbredu matki na przeżywalność cieląt żubra (*Bison bonasus* L.). Warszawa: Wydawnictwo SGGW; 2003

[9] Yon L, Duff JP, Ågren EO, Erdélyi K, Ferroglio E, Godfroid J, et al. Recent changes in infectious diseases in European wildlife. Journal of Wildlife Diseases. 2019;**55**:3-43. DOI: 10.7589/2017-07-172

[10] Karamon J, Larska M, Jasik A, Sell B. First report of the giant liver fluke (Fascioloides magna) infection in farmed fallow deer (*Dama dama*) in Poland—Pathomorphological changes and molecular identification. Bulletin of the Veterinary Institute in Pulawy. 2015;**59**:339-344. DOI: 10.1515/bvip-2015-0050

[11] Krzysiak MK, Iwaniak W, Kęsik–Maliszewska J, Olech W, Larska M. Serological study of exposure to selected arthropod–borne pathogens in European bison (*Bison bonasus*) in Poland. Transboundary and Emerging Diseases. 2016;**64**:1411-1423. DOI: 10.1111/tbed.12524

[12] Larska M, Wernery U, Kinne J, Schuster R, Alexandersen G, Alexandersen S. Differences in the susceptibility of dromedary and Bactrian camels to foot-and-mouth disease virus. Epidemiology and Infection. 2009;**137**:549-554. DOI: 10.1017/S0950268808001088

[13] Staśkiewicz R. Enzoocja pryszczycy u bizonów u żubrobizonów w zwierzyńcu w Smardzewicach w r. 1938. Medycyna Weterynaryjna. 1946;**2**:513-515

[14] Wróblewski K. Żubr Puszczy Białowieskiej. Poznań: ZOO Garden; 1927. 232 p

[15] Kita J, Anusz K. Infectious diseases in *Bison bonasus* in the years 1910-1988. In: Anusz KJ, K., editors. Health threats for the European Bison particularly in free-roaming populations in Poland.

Warszawa: The SGGW Publishers; 2006. pp. 17-26

[16] Kędrak-Jabłońska A, Budniak S, Szczawińska A, Reksa M, Krupa M, Krzysiak M, et al. Isolation and identification of *Pasteurella multocida* from European bison in Poland. Postępy Mikrobiologii. 2017;**56**:69-70

[17] Jaczewski Z. Spostrzeżenia z zakresu opieki weterynaryjnej w rezerwatach żubrów w latach 1952-1954. Roczniki Nauk Rolniczych. 1959;**E69**:297

[18] Kita J, Anusz K. Health threats for the European Bison particularly in free-roaming populations in Poland. Warszawa: SGGW Publishers; 2006. 319 p

[19] Podgurniak Z. Pathological lesions in the European bison caused by foot-and-moth disease in Polish reservations. Acta Theriologica. 1967;**30**:445-452

[20] Ciecierski H, Anusz K, Borko K, Anusz Z, Tsakalidis S. Occurrence of antibodies against Coxiella burnetii in wild animals in the foci of Q fever in 1985-1988. Medycyna Weterynaryjna. 1988;**44**:652-654

[21] Szarek J, Rotkiewicz T, Anusz Z, Khan MZ, Chishti MA. Pathomorphological studies in European bison (*Bison bonasus* Linnaeus, 1758) with seropositive reaction to *Coxiella burnetii*. Zentralblatt für Veterinärmedizin. Reihe B. 1994;**41**:618-624

[22] Kita J, Dziąba K, Piusiński W, Anusz K, Lenartowicz Z, Kowalski B, et al. A disease of genital organs of free-roaming male European bison in the Białowieża Primeval Forest (Poland). Medycyna Weterynaryjna. 1990;**46**:474-476

[23] Kita J, Anusz K, Zaleska M, Malicka E, Bielecki W, Osińska B, et al. Relationships among ecology, demography and diseases of European bison (*Bison bonasus*). Polish Journal of Veterinary Sciences. 2003;**6**:261-266

[24] Rzewuska M, Stefańska I, Osińska B, Kizerwetter-Świda M, Chrobak D, Kaba J, et al. Phenotypic characteristics and virulence genotypes of *Trueperella* (*Arcanobacterium*) *pyogenes* strains isolated from European bison (*Bison bonasus*). Veterinary Microbiology. 2012;**160**:69-76

[25] Borchers K, Brackmann J, Wolf O, Rudolph M, Glatzel P, Krasinska M, et al. Virologic investigations of free-living European bison (*Bison bonasus*) from the Bialowieza Primeval Forest, Poland. Journal of Wildlife Diseases. 2002;**38**:533-538

[26] Oleński K, Tokarska M, Hering DM, Puckowska P, Ruśść A, Pertoldi C, et al. Genome-wide association study for posthitis in the free-living population of European bison (*Bison bonasus*). Biology Direct. 2015;**10**:1-9. DOI: 10.1186/s13062-014-0033-6

[27] Radwan J, Kawałko A, Wójcik JM, Babik W. MHC-DRB3 variation in a free-living population of the European bison, *Bison bonasus*. Molecular Ecology. 2007;**16**:531-540

[28] Żórawski C, Lipiec M. Infekcja *Mycobacterium bovis* u żubra. Medycyna Weterynaryjna. 1998;**54**:178-180

[29] Żórawski C, Lipiec M. Przypadek uogólnionej gruźlicy u żubra. Medycyna Weterynaryjna. 1997;**53**:90-92

[30] Krajewska M, Welz M, Brewczyński P, Orłowska B, Anusz K. Gruźlica bydlęca w bieszczadzkiej populacji żubrów. Życie Weterynaryjne. 2014;**89**:148-151

[31] Krajewska M, Zabost A, Welz M, Lipiec M, Orłowska B, Anusz K,

et al. Transmission of *Mycobacterium caprae* in a herd of European bison in the Bieszczady Mountains, Southern Poland. European Journal of Wildlife Research. 2015;**61**:429-433

[32] Gortazar C, Diez-Delgado I, Barasona JA, Vicente J, De La Fuente J, Boadella M. The wild side of disease control at the wildlife-livestock-human interface: A review. Frontiers in Veterinary Science. 2015;**14**:27. DOI: 10.3389/fvets.2014.00027

[33] Krajewska M, Orłowska B, Anusz K, Welz M, Bielecki W, Wojciechowska M, et al. Gruźlica bydlęca w hodowli żubrów w Smardzewicach. Życie Weterynaryjne. 2016;**91**:50-53

[34] Welz M, Anusz K, Salwa A, Zaleska M, Bielecki W, Osińska B, et al. Gruźlica u żubrów w Bieszczadach. Medycyna Weterynaryjna. 2005;**61**:441-444

[35] Krajewska-Wędzina M, Kozińska M, Orłowska B, Weiner M, Szulowski K, Augustynowicz-Kopeć E, et al. Molecular characterisation of *Mycobacterium caprae* strains isolated in Poland. The Veterinary Record. 2018;**182**:292. DOI: 10.1136/vr.104363

[36] Krajewska-Wędzina M, Olech W, Kozińska M, Augustynowicz-Kopeć E, Weiner M, Szulowski K. Bovine tuberculosis outbreak in farmed American bison (*Bison bison*) in Poland. Polish Journal of Veterinary Sciences. 2017;**20**:819-821. DOI: 10.1515/pjvs-2017-0103

[37] Krasińska M, Krasiński ZA. Composition, group size and spatial distribution of European bison bulls in Białowieża Forest. Acta Theriologica. 1995;**40**:1-21

[38] Camp MJ, Rachlowa JL, Cisnerosc R, Roond D, Camp RJ. Evaluation of Global Positioning System telemetry collar performance in the tropical Andes of southern Ecuador. Natureza & Concervacao. 2016;**14**:128-131. DOI: 10.1016/j.ncon.2016.07.002

[39] Rhyan JC, Spraker TR. Emergence of diseases from wildlife reservoirs. Veterinary Pathology. 2010;**47**:34-39. DOI: 10.1177/0300985809354466

[40] Buttke DE, Decker DJ, Wild MA. The role of one health in wildlife conservation: A challenge and opportunity. Journal of Wildlife Diseases. 2015;**51**:1-8. DOI: 10.7589/2014-01-004

[41] Cunningham AA, Daszak P, Wood JLN. One Health, emerging infectious diseases and wildlife: Two decades of progress? Philosophical Transactions of the Royal Society of London. Series B, Biological Sciences. 2017;**19, 372**(1725). DOI: 10.1098/rstb.2016.0167

[42] Bolzoni L, De Leo GA. Unexpected consequences of culling on the radication of wildlife diseases: The role of virulence evolution. The American Naturalist. 2013;**181**:301-313. DOI: 10.1086/669154

[43] OIE. Guidelines for Wildlife Disease Surveillance: An Overview [Internet] 2015. Available from: http://www.oie.int/fileadmin/Home/eng/Internationa_Standard_Setting/docs/pdf/WGWildlife/OIE_Guidance_Wildlife_Surveillance_Feb2015.pdf [Accessed: Dec 18, 2018]

[44] Piusiński W, Malicka E, Bielecki W, Osińska B, Lenartowicz-Kubrat Z. Pathomorphological lesions in bisons in the Białowieża forest. Medycyna Weterynaryjna. 1996;**52**:386-388

[45] Krzysiak MK, Demiaszkiewicz AW, Pyziel AM, Larska M. Monitoring parazytologiczny żubrów w rezerwatach hodowlanych Białowieskiego Parku Narodowego. Medycyna Weterynaryjna. 2015;**71**(12):791-795

[46] Krzysiak MK, Jabłoński A, Iwaniak W, Krajewska M, Kęsik-Maliszewska J,

Larska M. Seroprevalence and risk factors for selected respiratory and reproductive tract pathogen exposure in European bison (*Bison bonasus*) in Poland. Veterinary Microbiology. 2018;**215**:57-65. DOI: 10.1016/j.vetmic.2018.01.005

[47] Salwa A, Anusz K, Arent Z, Paprocka G, Kita J. Seroprevalence of selected viral and bacterial pathogens in free-ranging European bison from the Białowieza Primeval Forest (Poland). Polish Journal of Veterinary Sciences. 2007;**10**:19-23

[48] Krasińska M, Krasiński ZA, Bunevich AN. Factors affecting the variability in home range size and distribution in European bison in the Polish and Belarussian parts of Białowieża Forest. Acta Theriologica. 2000;**45**:321-334

[49] Krasiński ZA. Dynamics and structure of the European bison population in the Bialowieża Primeval Forest. Acta Theriologica. 1978;**23**:3-48

[50] Thiede S, Spergser J, Rosengarten R, Jakob W, Streich WJ, Krasińska M, et al. Antibodies against *Mycoplasma bovigenitalium* in free-living European bison (*Bison bonasus*) with balanoposthitis. Journal of Wildlife Diseases. 2002;**38**:760-763

[51] Krzysiak MK, Dudek K, Krajewska M, Bednarek D, Szulowski K. Serological studies to determine the occurrence of Johne's disease and mycoplasma infection in the Northern-East Polish population of European bison (*Bison bonasus*). Polish Journal of Veterinary Sciences. 2014;**17**:721-723

[52] Highland MA, Herndon DR, Bender SC, Hansen L, Gerlach RF, Beckmen B. Mycoplasma ovipneumoniae in wildlife species beyond subfamily Caprinae. Emerging Infectious Diseases. 2018;**24**(12):2384-2386. DOI: 0.3201/eid2412.180632

[53] Rothschild BM, Martin LD, Lev G, Bercovier H, Bar-Gal GK, Greenblatt C, et al. Mycobacterium tuberculosis complex DNA from an extinct bison dated 17,000 years before the present. Clinical Infectious Diseases. 2001;**33**:305-311

[54] Didkowska A, Witkowski L, Orłowska B, Krzysiak M, Bruczyńska M, Krajewska-Wędzina M, et al. Improvement of ante-mortem diagnostics of bovine tuberculosis in European bison (*Bison bonasus*). In: Abstract Book. Conference 'European Bison in San Valley', 5-6 September 2018; Muczne. Warszawa: SMŻ; 2018. pp. 14-15

[55] Kohl TA, Utpatel C, Niemann S, Moser I. *Mycobacterium bovis* persistence in two different captive wild animal populations in Germany: A longitudinal molecular epidemiological study revealing pathogen transmission by whole-genome sequencing. Journal of Clinical Microbiology. 2018;**56**(9):e00302-18. DOI: 10.1128/JCM.00302-18

[56] Schmidbauer SM, Wohlsein P, Kirpal G, Beineke A, Müller G, Müller H, et al. Outbreak of *Mycobacterium bovis* infection in a wild animal park. The Veterinary Record. 2007;**161**:304-307

[57] Zimpel CK, Brum JS, de Souza Filho AF, Biondo AW, Perotta JH, Dib CC, et al. *Mycobacterium bovis* in a European bison (*Bison bonasus*) raises concerns about tuberculosis in Brazilian captive wildlife populations: A case report. BMC Research Notes. 2017;**10**:91. DOI: 10.1186/s13104-017-2413-3

[58] Kita J, Anusz K. Serologic survey for bovine pathogens in free-ranging European bison from Poland. Journal of Wildlife Diseases. 1991;**27**:16-20

[59] Majewska AC, Werner A, Cabaj W, Moskwa B. The first report of

Toxoplasma gondii antibodies in free-living European bison (*Bison bonasus bonasus* Linnaeus). Folia Parasitologica (Praha). 2014;**61**:18-20

[60] Moskwa B, Bień J, Kornacka A, Cybulska A, Goździk K, Krzysiak MK, et al. First *Toxoplasma gondii* isolate from an aborted foetus of European bison (*Bison bonasus bonasus* L.). Parasitology Research. 2017;**116**:2457-2461. DOI: 10.1007/s00436-017-5549-0

[61] Bień J, Moskwa B, Cabaj W. In vitro isolation and identification of the first *Neospora caninum* isolate from European bison (*Bison bonasus bonasus* L.). Veterinary Parasitology. 2010;**173**:200-205. DOI: 10.1016/j.vetpar.2010.06.038

[62] Straver PJ, van Bekkum JG. Isolation of malignant catarrhal fever virus from a European bison (*Bos bonasus*) in a zoological garden. Research in Veterinary Science. 1979;**26**:165-171

[63] Hänichen T, Reid HW, Wiesner H, Hermanns W. Malignant catarrhal fever in zoo ruminants. Tierärztliche Praxis. Ausgabe G, Grosstiere/Nutztiere. 1998;**26**:294-300

[64] Campolo M, Lucente MS, Mari V, Elia G, Tinelli A, Laricchiuta P, et al. Malignant catarrhal fever in a captive American bison (*Bison bison*) in Italy. Journal of Veterinary Diagnostic Investigation. 2008;**20**:843-846

[65] Li H, Taus NS, Jones C, Murphy B, Evermann JF, Crawford TB. A devastating outbreak of malignant catarrhal fever in a bison feedlot. Journal of Veterinary Diagnostic Investigation. 2006;**18**:119-123

[66] Martin C, Pastoret PP, Brocier B, Humblet MF, Saegerman C. A survey of transmission of infectious diseases/infections between wild and domestic ungulates in Europe. Veterinary Research. 2011;**42**:70

[67] Mac Lachlan JN, Guthrie AJ. Re-emergence of bluetongue, African horse sickness, and other orbivirus diseases. Veterinary Research. 2010;**41**:35

[68] Savini G, Afonso A, Mellor P, Aradaib I, Yadin H, Sanaa M, et al. Epizootic haemorrhagic disease. Research in Veterinary Science. 2011;**91**:1-17

[69] Glunz R. Bluetongue disease at European bison—Symptoms, section results, treatment, vaccination. In: Proceedings of VI Conference 'Bison in the Network Natura 2000,' European Bison Friends Society, Cisna, Poland, 15-16 September. 2008. pp. 18-19

[70] Orłowska A, Trębas P, Smreczak M, Marzec A, Żmudziński JF. First detection of bluetongue virus serotype 14 in Poland. Archives of Virology. 2016;**161**:1969-1972. DOI: 10.1007/s00705-016-2857-0

[71] European Commission. VBM Molecular Epidemiology Report. 2012. Available from: https://ec.europa.eu/food/sites/food/files/animals/docs/ad_control-measures_bt_molec-epidemiology_report_20121130_poland.pdf [Accessed: Dec 21, 2018]

[72] Ruiz-Fons F, Sánchez-Matamoros A, Gortázar C, Sánchez-Vizcaíno JM. The role of wildlife in bluetongue virus maintenance in Europe: Lessons learned after the natural infection in Spain. Virus Research. 2014;**182**:50-58. DOI: 10.1016/j.virusres.2013.12.031

[73] Kęsik-Maliszewska J, Krzysiak MK, Grochowska M, Lechowski L, Chase C, Larska M. Epidemiology of Schmallenberg virus in European bison (*Bison bonasus*) in Poland. Journal of Wildlife Diseases. 2018;**54**:272-282. DOI: 10.7589/2017-07-159

[74] Hoffmann B, Scheuch M, Höper D, Jungblut R, Holsteg M, Schirrmeier H,

et al. Novel orthobunyavirus in Cattle, Europe, 2011. Emerging Infectious Diseases. 2012;**18**:469-472

[75] Larska M, Kęsik-Maliszewska J, Kuta A. Spread of Schmallenberg virus infections in the ruminants in Poland between 2012 and 2013. Bulletin of the Veterinary Institute in Pulawy. 2014;**58**:169-176. DOI: 10.2478/bvip-2014-0026

[76] Larska M, Krzysiak M, Smreczak M, Polak MP, Żmudziński JF. First detection of Schmallenberg virus in elk (*Alces alces*) indicating infection of wildlife in Białowieża National Park in Poland. Veterinary Journal. 2013;**198**(1):279-281. DOI: 10.1016/j.tvjl.2013.08.013

[77] Tarlinton RE, Daly JM, Dunham SP, Kydd JM. Schmallenberg virus: Could wildlife reservoirs threaten domestic livestock? Veterinary Journal. 2013;**198**:309-310

[78] Larska M, Krzysiak MK, Kęsik-Maliszewska J, Rola J. Cross-sectional study of Schmallenberg virus seroprevalence in wild ruminants in Poland at the end of the vector season of 2013. BMC Veterinary Research. 2014;**10**:967. DOI: 10.1186/s12917-014-0307-3

[79] Koch HG, Axtell RC. Attraction of C*ulicoides furens* and *C. hollensis* (Diptera: Ceratopogonidae) to animal hosts in a salt marsh habitat. Journal of Medical Entomology. 1979;**15**:494-499

[80] Zimmer JY, Verheggen FJ, Haubruge E, Francis F. Orientation behaviour of *Culicoides obsoletus* (Diptera: Ceratopogonidae), a relevant virus vector in northern Europe, toward host-associated odorant cues. Veterinary Parasitology. 2015;**211**:274-282

[81] Adaszek L, Dzięgiel B, Krzysiak M, Skrzypczak M, Adaszek M, Staniec M, et al. Detection of Borrelia burgdorferi sensu lato DNA in the blood of wild bison from Białowieza primeval forest in eastern Poland. Polish Journal of Veterinary Sciences. 2014;**17**:713-715

[82] Siński E, Gill J, Rijpkema S. Prevalence of antibodies to *Borrelia burgdorferi* in European bison (*Bison bonasus*) from Białowieza Primeval Forest. Roczniki Akademii Medycznej w Białymstoku. 1996;**41**:111-116

[83] Biernat B, Karbowiak G, Stańczak J, Masny A, Werszko J. The first detection of the tick-borne encephalitis virus (TBEV) RNA in *Dermacentor reticulatus* ticks collected from the lowland European bison (*Bison bonasus bonasus* L.). Acta Parasitologica. 2016;**61**:130-135. DOI: 10.1515/ap-2016-0017

[84] Biernat B, Karbowiak G. Study on the occurrence of tick-borne encephalitis virus RNA in European bison (*Bison bonasus*) eliminated at Białowieza Primeval Forest (north-eastern Poland) in 2005-2009. Annals of parasitology. 2014;**60**:99-102

[85] Karbowiak G, Víchová B, Werszko J, Demiaszkiewicz AW, Pyziel AM, Sytykiewicz H, et al. The infection of reintroduced ruminants—*Bison bonas*us and *Alces alces*—With *Anaplasma phagocytophilum* in northern Poland. Acta Parasitologica. 2015;**60**:645-648. DOI: 10.1515/ap-2015-0091

[86] Matsumoto K, Grzeszczuk A, Brouqui P, Raoult D. *Rickettsia raoultii* and *Anaplasma phagocytophilum* in *Dermacentor reticulatus* ticks collected from Bialowieza Primeval Forest European bison (*Bison bonasus bonasus*), Poland. Clinical Microbiology and Infection. 2009;**15**(Suppl 2):286-287. DOI: 10.1111/j.1469-0691.2008.02238.x

[87] Larska M, Krzysiak MK, Jabłoński A, Kęsik J, Bednarski M, Rola J. Hepatitis E virus antibody prevalence in wildlife in Poland. Zoonoses and Public Health. 2015;**62**:105-110. DOI: 10.1111/zph.12113

Section 3

Wildlife and Human Interaction

Chapter 8

An Assessment of the Human-Wildlife Conflict across Africa

Benjamin-Fink Nicole

Abstract

The coexistence between humans and mammals across Africa has led to Human Wildlife Conflict (HWC) due to the competition for limited natural resources. Over the past two decades, I have focused my research on conservation issues that either resulted from or induce human-wildlife conflict. Conflicts are intensified in regions where dense human populations live in close proximity to nature, and where livestock holdings and crop fields form a significant part of rural livelihoods. As a result, both people and wildlife suffer tangible consequences; therefore, creating the need for stakeholder's involvement and their willingness to adopt conservation-based behaviors, as key ingredients for feasible and effective conservation counter measures. This chapter provides a comprehensive review of the wide array of drivers and conservation implications of HWC incidences throughout Africa. An in-depth analysis is essential to understanding the problem and support future conservation prospects. Examples explore key case studies ranging from decreasing numbers of the charismatic forest dwelling elephant (*Loxodonta cyclotis*) in the DRC, to increasing numbers of waterbuck (*Kobus ellipsiprymnus*) in Mozambique, and varying numbers of lion populations bordering Kruger National Park in South Africa. Concluding with conflict resolution strategies employed across Africa and recommendations for the effective conservation of the world's most endangered mammals.

Keywords: Africa, biodiversity, carnivores, community based conservation, ecology, human-wildlife conflict, mammals, poaching, restoration, strategic planning, transfrontier parks

1. Introduction

Conflicts between people and mammals currently rank among the highest main threats to conservation in Africa. The World Conservation Union [1] defines Human-Wildlife Conflict (HWC) as a threat resulting directly from the competition between rural communities and wild animals over natural resources, entailing consequential tangible costs to both people and wildlife. Alarmingly, HWC primarily involves endangered species and embeds broader environmental impacts on ecosystem equilibrium and biodiversity preservation. The most common HWC incidences undermine human welfare, health and safety; which result in zoonotic diseases, physical injury, and loss of human lives. Sustainability is rooted in the continuous loss of economic revenue due to the damage to property and infrastructure (e.g. agricultural crops, water installation, etc.), livestock depredation, and the transmission of domestic animal diseases (**Figure 1**). In fact, the meat export

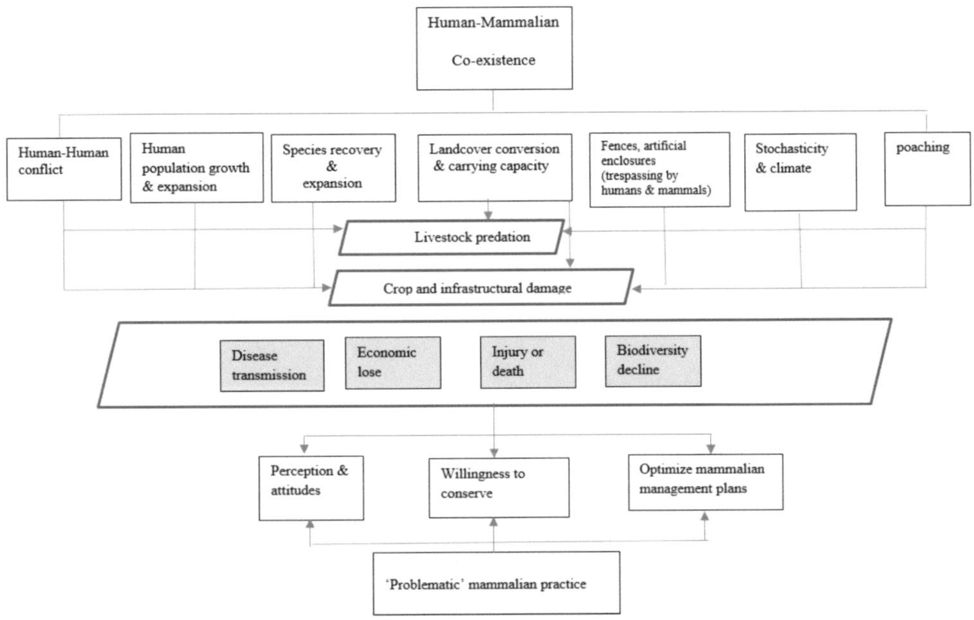

Figure 1.
The connection between human-mammalian co-existence and its drivers for conflict. Casual interlinkages are illustrated by directional arrows. Consequential variables are highlighted with slanted frames and in gray boxes.

market sector of South Africa came to a halt in 2008 due to concerns over a possible outbreak of Hand, Foot, and Mouth Disease (HFMD), known to be spread by ungulates such as wildebeest.

2. Overview of HWC drivers across Africa

Over the past two decades, I have focused my research on the implications of human-wildlife conflict interactions worldwide. In Africa, HWC primarily occurs in regions where rural communities live adjacent to protected areas. This chapter explores drivers for, and solutions to, regional case studies throughout Africa. Much of these drivers enable significant shifts in the dynamics of endangered mammalian species (i.e., abundance, composition, and distribution of prey and predator population). The conservation concern arises from the developed negative attitudes of the local people toward wildlife causing exacerbated and dejection of the conservations' efforts. In order to break this threatening cycle, there is a need to protect rural livelihoods, reduce their vulnerability, and counterbalance losses with benefits and foster community-willingness to participate in conservation.

2.1 Human population growth and land conversion

Human population growth inherently entails settlement expansion into and around protected areas [2]. Additionally, more often than not such expansion entails land cover conversion and shifts in wild herbivore and predator dynamics.

Human population growth entails settlement expansion which has resulted in the infringement into mammalian habitats, constriction of species into marginal habitat patches, and increased direct competition for coexistence. Human encroachment

DOI: http://dx.doi.org/10.5772/intechopen.82793

into national reserves is documented to occur in Nigeria, Uganda, Ghana, Congo, Kenya, Tanzania, Zanzibar, and South Africa to name a few countries.

As human population grows and settlement expands, land cover conversion is driven by the need for economic profitability and livelihood sustainability. Research done in Samburu, Trans-Mara, Taita, and Kwale in Kenya illuminate the interlinkage between intensified conflict, land conversion and the development of small-scale farming [3]. Provide details!! Over the last 4 decades, state and trust ranches have been subdivided and sold as smallholdings and commercially cultivated horticultural areas. The unintended result is an artificially improved carrying capacity which alters mammalian ecology. These crops may serve as favorable feeding grounds for non- residential predators, thereby altering the natural distribution and ecology of predators and the dynamics of the prey population abundances. Need to add more details!!!

Growing densities in livestock populations can create an overlap of diets and forage competition with wild herbivores, resulting in overgrazing and decline or even local extinction in wild herbivore populations. As such, livestock becomes an important source of prey for predators. As a response, landowners often shoot and kill predators found on their land, or on the neighboring lands in order to avoid economic loss associated with livestock predation.

2.2 HWC results in monetary loss

Villagers living adjacent to Tasavo Conservation Area (TCA), Kenya, indicate the severity of the ongoing conflict; 64.8% of respondents indicated that their agriculture was destroyed, 28.8% reported livestock predation, 4% indicated damage to water structures, and 2.3% reported damage to fences (add reference).

A variety of mammals have been documented to cause crop damage, such as primates, rhino, elephant, deer, and wild pig. Crop loss includes mainly banana, cassava, maize, and sweet potatoes, millet, and cotton. In fact, the endangered African bush elephant (*Loxodonta africana*) accounted for 4–7% of annual crop loss in Kibale national Park, Uganda [2]. The park borders the communities, which is 54% of cultivated land within 1Km of Kibale's border. Landowners report a loss of maize, cassava, sweet potato, and banana due to elephants, bushpig (*Potamochoerus sp.*), olive baboons (*Papio cynocephalus*), and redtail monkey (*Cercopithecus ascanius*). Similarly, farmers consider most medium and large-size mammals as a threat to their crops in Jozani Forest Reserve located on Unguja, Zanzibar. In particular, loss of coconut harvest is attributed to one of the most endangered primates, the red colobus (*Procolobus kirkii*).

In Caprivi, Namibia, the largest free ranging population of elephants (5000) shares their habitat with local people. Elephants dispersing from the two Caprivi National Parks are documented to be responsible for a lot of crop damage although the financial losses are less than lion (*Panthera leo*) attacks on livestock (US$39,200, US$85,570, respectively) [4].

For many communities across Africa, the cost of coexisting with wildlife is the loss of valuable revenue due to livestock predation by the endangered Leopard (*Panthera pardus*), the vulnerable Lion and Cheetah (*Acinonyx jubatus*), or the Spotted hyenas (*Crocuta crocuta*).

Commercial cattle ranches supply meat to both national and international markets. Such ranches located adjacent to the Tsavo East National Park in Kenya suffer losses of large domestic animals (i.e., cows, bulls, and steers) due to predation by lions and spotted hyenas, in addition to losses of small domesticated animals (i.e., young cattle) following cheetah attacks [5]. In a 4 year study, ranchers lost annually on average 2.4% of their herd, accounting for 2.6% of their economic revenue (US$8749).

Similarly, traditional agro-pastoralism is prone to livestock deprivation in Zimbabwe. During 1993–1996, 241 livestock were reported as killed by lions, leopards, and yellow baboons (*Papio anubis*), accounting for 34, 12, and 54% of the kills respectively [6]. These amounts account for an annual loss of 142% of total family's income per household. These mammals differ in their targeted prey and therefore, in the economic damage; while the large predators kill cattle and donkeys, baboons kill smaller livestock (e.g., sheep and goats). As a result, lions account for a large share of the economic loss.

2.3 Fencing and artificial enclosures

Across Africa, fences provide a spatial barrier around game reserves, conservancies, and homestead areas.

Populations are closed, finite and natural processes do not take place (e.g., dispersal, emigration, and colonization dynamics). This not clear!! Negative consequences may be in one of the following forms: hybridization and introgression, genetic bottlenecks, inbreeding or outbreeding depression, and shifts in local or regional ecological equilibrium. As a result of such artificial enclosures, game ranch owners are faced with the need to intensely manage populations on their land while weighting economic profitability with genetic concerns stemming from small and closed populations (e.g., inbreeding, outbreeding depression, and bottlenecks) [7]. As such, increasing risks posed to endangered mammals.

Conflicts over natural resources have increased as communities, particularly pastoralists, compete for diminishing water, pasture and food resources with wildlife. Tasavo Conservation Area (TCA) is located adjacent to traditional villages in Kenya where traditional pastoralism is practiced. Despite fences, people trespass with their livestock into the TCA so that they may access resources that are lacking (e.g., grazing). Similarly, Maasai herdsmen, a semi-nomadic pastoralist group, drive their animals into the Maasai Mara game in Kenya on an ongoing basis [8].

The effectiveness of natural or man-made fences depends on behavior of different mammalian species. For example, burrowing animals such as bushpig can breach such barriers and enable additional mammalian species to do so (such as lions utilizing their holes around Kruger National Park in South Africa). Similarly, baboons, lions and leopards have been documented to jump over fences and into livestock enclosures at the Wildlife Research Area in Zimbabwe [9]. In Laikipia District situated in Northern Kenya, pastoralists discovered that domestic animals experience a lower depredation rate when penned in corals overnight (when carnivore tend to hunt) [10].

2.4 Species' localized recovery

Conservation success is often synonymous with the successful recovery of declining endangered populations through effective management strategies. However, it is worth noting that the social organization, habitat and prey requirements, and home-range distributions of such recovered populations often serve as a renewed source for HWC. Examples include: lions straying out of the Kruger Park Reserve in South Africa and into adjacent villages, and bull elephants on musth acting aggressively and venturing into local communities across Kenya. Recent recovery programs have additionally contributed to the recolonization by cheetahs of their original home range including rural areas in Botswana; and in the process have increased the frequency of domestic livestock predation.

2.5 Climate factors and stochastic events

Climate is a determining factor in mammalian colonization (i.e., home range and distribution expansion, contraction, or shift). As climate chance becomes increasingly rapid, people can expect an increase in HWC incidences. Concurrently, seasonal changes in rainfall are directly correlated with lion predation intensity on livestock in Tsavo National Park in Kenya [5]. In fact, monthly rainfall predicts lion attacks on livestock during seasonal rains. Similarly, the intensity of livestock predation by predators increased in Sengwa Wildlife Research Area, Zimbabwe, during the dry season when vegetation cover is scarce [6].

2.6 Emerging markets

South Africa experienced an annual increase of 5.6% in land used for wildlife ranching during 1991–2000 [7]. Diminished economic profitability of livestock, coupled with an increase in stock theft, resulted in a shift from traditional agriculture and cattle farming to ungulate ranching within the private ranching industry. With an annual return on investment per hectare higher than any agriculturally based market (upwards of 80%), the game ranching industry generates annually R4.7 billion [11, 12].

More recently, the market for ungulate color variance has evolved in South Africa. Proactively breeding for morphological characteristics (i.e., coat color) is a relatively new avenue for economic revenue, yielding upwards of 1 billion Rand/year. This sector is emerging in its scope and has the capacity to produce 50–1000% ROIs in full. A primary example is the breeding of a bull golden wildebeest (*Connochaetes*) with a cow blue wildebeest (*Connochaetes taurinus*); the offspring is referred to as "Split Golden" and were sold in private auctions during 2014 for 513, 137 rand (a 541% increase in selling price compared to 80,000 rands in 2004) [7]. These Split Golden wildebeests are then mated with golden bulls (i.e., backcrossed) to facilitate the expression of the recessive allele for a golden color coat. Similarly, the price of offspring that result from breeding a blue wildebeest with split king wildebeest increased 10-fold during 2012–2014 (81,553–882,917 rands, respectively). The average price of a pure breed blue wildebeest in 2014 was 3626 rands. Breeding for color variance may also occur by selectively breeding blue wildebeest with split golden wildebeest or blue wildebeest with king wildebeest. The high ROIs dictate mammalian artificial overlap and stocking within the confinement of these reserves; thus, in turn, altering intra and interspecies dynamics, resulting in the potential of significant increases in HWC frequencies.

2.7 The illegal wildlife trade (IWT)

The illegal wildlife trade can be clustered into one of two categories: (1) the bushmeat trade, the illegal harvest of non-domesticated animals in tropical forests in order to meet local and regional consumption demand, and (2) the international trade in body parts of endangered animals to meet the international black-market demands [9, 10]. The first primarily consists of hunting ungulates and, the second entails the poaching of rhinos, elephant, and pangolins to name the most popular. It is important to note that often times, poaching entails the illegal harvest of ungulates which are not classified as endangered including Kudu (*Tragelaphus*) in Mozambique and waterbuck (*Kobus*) in Zimbabwe and Mozambique.

Bushmeat represents an important source of animal protein for humans in tropical Africa. In Bénoué National Park (BNP), Cameron, and Sam-Sam villages, Ghana, is the local diet, whereby bushmeat consists of up to 24% of animal protein intake. Two primary negative impacts are associated with bushmeat consumption; the transfer of Zoonotic diseases and biodiversity loss. The Ebola outbreak crisis across West Africa in 2016 resulted in a decreased consumption rate in high-income households than in low-income households [16]. However, daily meal frequency decreased during the crisis, while the diversity of food items and preferences for bushmeat species remained constant. In Garamba National Park and neighboring Azande Hunting Reserve, Congo, bushmeat hunting of primates increased fivefold during political instability and armed conflict resulting from warfare.

The Illegal Wildlife Trade (IWT) refers to the trade in animal body parts among international borders. With an annual circulation of $17 billion annually, IWT is driven by the consumer belief system that consuming an animal or its body parts, then one resumes the animals' potency [14]. Africa is home to the most endangered mammalian population supplies. The most trafficked animal worldwide is the pangolin (*Manis*) for its scales, the white rhino (*Ceratotherium simum simum*) and the black rhino (*Diceros bicornis*) for their horn, and the elephant for its tusks. In 2011, South Africa was home to 83% of Africa's rhinos and nearly three-quarters of all wild rhinos worldwide. Alarmingly, on average, one rhino was poached every 8 hours during 2015. Specifically, three South African provinces (Limpopo, North West, and KwaZulu-Natal) are home to 90% of South Africa's rhino population, experiencing 75% of all poaching during 2012–2017.

2.8 Civil war and poaching

During the civil war and Human-Human Conflict (HHC) in South Sudan and The Democratic Republic of the Congo elephant populations significantly decreased. Their decreased numbers were attributed to increased poaching and dispersal. Studies conducted during political stable times indicate that elephant populations re-established their local home-ranges after HHC. Provide details!!!

3. Conservation strategies designed to address increasing HWC

Most HWC incidences may be considered *wicked* conservation problems; their multi-complexity is rooted in the interlinkage imbedded in social, economic, biological, and stochastic variables (**Figure 2**, [13]). They vary according to different taxonomic groups within the Mammalia class and sub-regions across the African continent. Their inherited uncertainty can be quantified by various technology and formulas, which can then be applied to assist local mammalian management [13]. Moving forward, conservation of mammals (and all other species) in Africa will only be affected if bottom-up and top-down approaches are implemented simultaneously. The following section offers an assessment of implemented strategies which mitigate and minimize HWC. These are subdivided into two: (1) ground level applications and (2) wide scale conceptual applications.

3.1 Ground level applications

The Maasai and Samburu local communities utilize homestead fencing whereby crop fields and livestock are fenced in corals and wildlife dispersal is enabled. Limited depredation is attributed to such established husbandry practices which utilize wire, acacia, wicker or solid enclosures (listed from the most effective to the least).

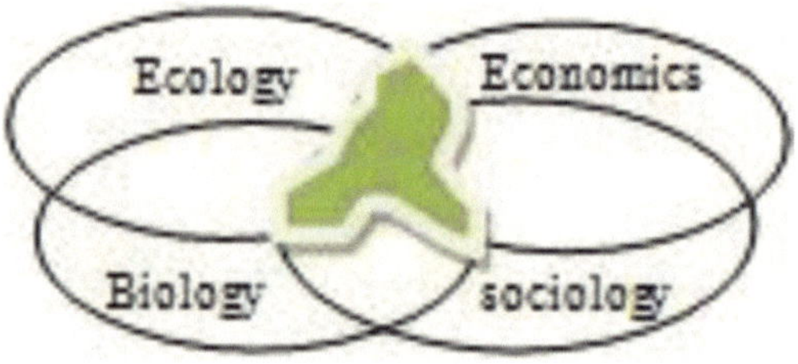

Figure 2.
The multifaceted human mammalian interactions and conflict drivers are addressed within a wicked conservation contextual problem (illustrated in green). Effective strategies to decrease HWC are found within the interlinked area. Adapted from Benjamin-Fink and Reilly [13].

In addition to these conventional fences, electric fencing such as those required in reserves that enclose the "BIG 5" (i.e., lion, buffalo (*Syncerus caffer*), leopard, elephant, and rhino), or constructing roofs on the fences, are viable alternatives for a more sophisticated solution. Electric fences have proven to be effective at creating a spatial separation between human settlements and mammals in Nyeri and Laikipia districts in Kenya encompass the Endarasha and Ol Moran villages. In fact, electric fences in the East Caprivi region of Namibia have successfully deterred elephants from destroying crops [4, 5]. Equally important, the return on investment in eletric fences ocurs within 5 years, which may then be reallocated to offset associated costs.

Crop damage and death incidences are primarily associated with male mammals, most of which are polygynous. This is due to their large body size and mass. As a result, culling a male elephant would have the equivalent crop economic result of removing a family of 20 elephants. The demography of a population is correlated to the female population size, age of first reproduction, litter size, interval between births, age of reproductive senescence, and mortality. In a polygynous species such as elephants, the male contributes sperm, providing no additional source of energy in the rearing of offspring. As such, a surplus of males is not uncommon. The critical adult sex ratio varies amongst mammalian species and is additionally correlated to spatial distribution, density, seasonality in breeding, duration and periodicity of oestrus in females, reproductive behavior, etc. With that in mind, poaching is primarily targeted toward male and female tuskers, which consequently skews the demography. As such, culling of problematic male elephants as a policy to reduce conflicts is not widely recommended.

The practice of translocation of "problematic" animals is a viable alternative to culling. However, when taken in an *ad hoc* fashion, this mammalian management practice may also have negative effects on the genetic pool; conservation programs should consider all aspects of ecological implications [7]. Ungulate translocations need to be considered as the possibility of inbreeding, outbreeding, and hybridization high risk implications may additionally be posed in the case of carnivores, particularly when territories are already occupied by individuals of the same species.

Herders are fearless when protecting their animals in East Africa. The task of deterring lions, hyenas, and cheetahs, is given to young boys or adult males. Undesired shooting and death of the animals themselves occurs frequently. The utilization of guard dogs as a safer tactic is employed across South Africa and Botswana, whereby the dog's bark is sufficient to scare predators away.

An intensified version of proactive protectors is anti-poaching units (APU) (see below).

3.2 Wide scale conceptual applications

A community based natural resource management scheme (CBNRMS) has been established in the Caprivi region of Namibia, where the eco-tourism industry and hunting concessions provide the lions a share of local economic revenue. The interlinkage of livelihoods with wildlife-based revenue (particularly large mammalian species and carnivores) is a scheme employed across Kenya and Namibia. An example of indirect economic income is the creation of employment opportunities via the tourist sector. Kibale National Park in Uganda foster positive attitudes toward the park by directly sharing a percent of generated revenues with local communities.

Moreover, Bénoué National Park (BNP), Cameroon, was established in 1968. Prior to which the land was owned by local villages. Concurrently, their revenue sources include crop, livestock, fishing, and hunting as such, the average loss per household is an estimated annual 31% of their economic revenue generated from crop and 18% of their economic revenue livestock generated from livestock due to elephants, baboons, and warthog (*Phacochoerus aethiopicus*) [15, 16]. The government currently subsidies their economic loss. In a nut shell, economic returns benefit rural communities and serve as incentives which motivate them to protect wildlife outside protected areas and to discourage poaching.

The role of anti-poaching units (APUs) is diverse. Their primary objective is to halt the poaching of charismatic mammals (e.g., pangolin, elephant, rhino, etc.`) and their activities range from aerial patrols conducted by satellites and helicopters, to ground patrols conducted by ground troops traveling by foot and/or in vehicles. Additionally, they serve as a vital peace-keeping governing body within local communities. At times, patrolling officers may extend themselves to assist surrounding villages in resolving common disputes surrounding criminal activities. Such processes and involvement facilitate a mutual trusting relationship and positive attitudes toward mammalian conservation.

Settlement of rights is a scheme by which the reimbursement for crop and livestock losses, or of loss of human life or injury, has a fixed quota of exploited commodities. It legitimizes the rights of local communities to resources within reserve zones. This, in turn, promotes positive attitudes and perceptions toward mammalian conservation and a higher tolerance for HWC incidences.

4. Recommendations and conclusions

This chapter highlights the importance of considering human-mammalian interactions when poising to understand the most critical conservation challenges facing biodiversity and formulating effective management strategies. It highlights common problems and solutions across biogeographical regions and demonstrates that conflicts have similar causes and effects, and that detailed information is the key to the development of appropriate strategies for resolving the problem and conserving different ecosystems and their inhabitants.

Many international organizations strive to implement mammalian conservation strategies that will decrease HWC incidences and bring back endangered species from the brink of extinction. Conservation beyond borders (CBB) provides an example of considering the most challenging conservation phenomenon within their complexity and formalizing comprehensive solutions.

The most sustainable approaches ensure the development of a local economy derived partially from wildlife (e.g., revenue from nature reserves), as well as a reduction in the vulnerability of rural communities to livestock

DOI: http://dx.doi.org/10.5772/intechopen.82793

depredation and crop damage by various mammalian species. In order to optimize conservation efforts, management plans should be based on scientific knowledge, combined with indigenous knowledge, practical local knowledge, and collaboration.

Acknowledgements

I would like to thank many African colleges who have contributed directly or indirectly to insights highlighted in this chapter. Their valuable time and knowledge was instrumental to the conceptual formulation of ideas. A special thanks to Neev, Sophy, and Emma for providing important guidance.

Conflict of interest

There are no conflicts of interest the author wishes to point out or declare.

Author details

Benjamin-Fink Nicole
Conservation Beyond Borders, Minnesota, USA

*Address all correspondence to: nicole@conservationbeyondborders.org

References

[1] Preventing and mitigating human-wildlife conflicts IUNC. World Park Congress; 2003. Available from: http://www.iucn.org/themes/wcpa/wpc2003/pdfs/outputs/recommendations/approved/english/pdf/r20.pdf

[2] Naughton-Treves L. Predicting patterns of crop damage by wildlife around Kibale National Park, Uganda. Conservation Biology. 1998;**12**(1):156-168

[3] Wanyingi NJ. Determinants of human-elephant conflicts in Shimba Hills ecosystem, Kenya [doctoral dissertation]. 2016. Available from: http://erepository.uonbi.ac.ke/bitstream/handle/11295/97529/betty%20final%20project.pdf?sequence=1&isAllowed=y

[4] O'Connell-Rodwell CE, Rodwell T, Rice M, Hart LA. Living with the modern conservation paradigm: Can agricultural communities co-exist with elephants? A five-year case study in East Caprivi, Namibia. Biological Conservation. 2000;**93**(3):381-391

[5] Patterson BD, Kasiki SM, Selempo E, Kays RW. Livestock predation by lions (*Panthera leo*) and other carnivores on ranches neighboring Tsavo National Parks, Kenya. Biological Conservation. 2004;**119**(4):507-516

[6] Butler JR. The economic costs of wildlife predation on livestock in Gokwe communal land, Zimbabwe. African Journal of Ecology. 2000;**38**(1):23-30

[7] Benjamin-Fink N, Reilly BK. Conservation implications of wildlife translocations; the state's ability to act as conservation units for wildebeest populations in South Africa. Global Ecology and Conservation. 2017;**12**:46-58

[8] Mutahi MD, Mwangi GP, Mutura BR. Impact of Electric Fence on Pastoralists Movement in Ol Pejeta Eco-System in Laikipia, Kenya; 2012

[9] Pennisi E. People are hunting primates, bats, and other mammals to extinction. Science. 2016

[10] Distefano E. Human-Wildlife Conflict Worldwide: Collection of Case Studies, Analysis of Management Strategies and Good Practices. Rome, Italy: Food and Agricultural Organization of the United Nations (FAO), Sustainable Agriculture and Rural Development Initiative (SARDI); 2005. Available from: FAO Corporate Document repository http://www.fao.org/documents

[11] Cloete PC, Van der Merwe P, Saayman M. Game Ranch Profitability in South Africa. ABSA; 2015. pp. 50-87

[12] Lindsey SL, Green MN, Bennett CL. The Okapi: Mysterious Animal of Congo-Zaire. University of Texas Press; 2010

[13] Benjamin-Fink N, Reilly BK. A road map for developing and applying object-oriented Bayesian networks to "WICKED" problems. Ecological Modelling. 2017;**360**:27-44

[14] IUCN: International Union for Conservation of Nature. Available from: https://www.iucn.org/

[15] Weladji RB, Tchamba MN. Conflict between people and protected areas within the Bénoué wildlife conservation area, North Cameroon. Oryx. 2003;**37**(1):72-79

[16] Granados A, Weladji RB. Human-elephant conflict around Bénoué National Park, Cameroon: Influence on local attitudes and implications for conservation. Human Dimensions of Wildlife. 2012;**17**(2):77-90

Chapter 9

Resident Hunting Ban in Serengeti District and Its Implications to People's Livelihood and Wildlife Population

Abiud L. Kaswamila and Augustino E. Mwakipesile

Abstract

In 2002 Serengeti District Council entered an agreement with Singita Grumeti Reserve Limited to ban resident hunting by compensating the District Council TZS 460 million (USD 200,000) per annum. This study assessed the implications of the ban on communities' livelihood and wildlife populations in the district. Up to 2011 about TZS 1.7 trillion (USD 727, 000) had been paid to the District Council. Findings reveal that communities were not involved in the decision and that the ban had mixed results on community livelihood. Communities mentioned provision of development infrastructures and students' scholarship and increase of game species as benefits emanating from the ban. Denial to access game meat, increased destructive wildlife particularly elephants, misallocation of funds by the Council, increased food insecurity, and killings of people by wildlife were perceived as costs. The study concludes that the ban has significantly boosted the Council's revenues and increased wildlife populations. However, communities have been denied their constitutional rights of hunting and access to cheap source of protein. Further, communities have witnessed increased human-elephant conflict and food insecurity. The study recommends involvement of communities in such major decisions affecting people's livelihood and the need for research before implementing such a decision.

Keywords: conservation, development, Singita Grumeti reserve, human-wildlife conflicts, human-wildlife interactions, ecosystems

1. Introduction

Around the world, resident hunting as part which is generally organized and managed by the state as a public service [1] has proven to be a powerful tool to promote conservation when conducted in ways that are biologically sound within appropriate governance and institutional settings [2]. License fees and taxes are set and taken by the state and in most situations are reinvested in managing the resource. In developing countries apart from conservation, public hunting can be regarded as a source of income as well as a livelihood strategy in terms of food security and in combating malnutrition. On the other hand, public hunting model has not worked effectively for conservation in Africa and other developing regions such as Central Asia and even Mexico [3]. For example, governance, accountability mechanisms, and state agencies are unable to capture and channel license or tax revenues to manage and support conservation programs as they do in the North American Model [3].

In Tanzania resident hunting is recognized by the wildlife policy [4] which states that it is the right of indigenous Tanzanians to have legal access to wildlife use, and different scales of fees for tourists and residents to hunt have been established through the enabling legislation. The country has over 140 hunting concessions covering an area in excess of 250,000 km^2 that are licensed to conduct both tourist hunting and resident hunting. These concessions are distributed throughout the country in either Game Reserves and Game Controlled Areas, open areas, or Wildlife Management Areas. Schedules of the Hunting Regulations that support the Wildlife Conservation Act specify types of animals that may or may not be hunted on a hunting license. A wide range of animals (approximately 60 species) can legally be hunted by tourist hunters in Tanzania. However, the giraffe, cheetah, rhino, and wild dog are protected game and cannot be hunted in Tanzania. The Hunting Regulations stipulate the fees for hunting every type of animal and further specify a minimum number of hunting days. Currently there are over 40 hunting outfitters leasing concessions in Tanzania.

In Serengeti District before the 2002 joint venture between Serengeti District Council (SDC) and Singita Grumeti Reserve Limited (SGRL) which banned resident hunting, the latter was undertaken using existing regulations. Resident hunting was formally carried out in Nyichoka and Sibora open areas of Serengeti District. The chapter assesses the implications of the hunting ban on local communities' livelihood and examines the wildlife population trend after the ban.

2. Study area

Serengeti District (10,373 km^2) is located on the Eastern part of Mara region (**Figure 1**). The district has 10 wildlife areas, viz., Serengeti National Park (7000 km^2), Grumeti Game Reserve (68.37 km^2), Ikorongo Game Reserve (189.68 km^2), IKONA Wildlife Management Area (WMA) (148 km^2), and open areas—IKONA Open Area, Issenye Open Area, Issenye Reminder, Nyakitono Reminder, Robanda Open Area, and Sasakwa, having a total of 2306 km^2. About 659 km^2 is the area for agriculture, livestock keeping, and for settlements.

Average rainfall in the area is about 700 mm per annum. The temperatures in the area show a relatively constant mean monthly maximum of 27–28°C. The minimum temperature varies from 16°C in the hot month of October–March to 13°C during May–August. The relief ranges from 1144 to 1380 m above sea level.

This study involved three communities of Natta Mbiso, Machochwe, and Park Nyigoti. The criteria for picking these communities were their traditional link to wildlife utilization and adjacency to former hunting areas and/or protected area networks in the district. As for game population, emphasis was on IKONA WMA, which entered an agreement with Singita Grumeti Reserve Limited (SGRL).

SGRL under a multi-billionaire businessman Paul Tudor II was registered in 2002 and has tourism right tenure over the Grumeti Game Reserve, Ikorongo Game Reserve, and Ikoma Wildlife Management Area. The SGRL which is one of the most ambitious conservation and tourism projects in Africa is a private game concession and has excellent year-round game viewing and is also ideally placed for annual wildebeest migration. Currently SGRL project operates three 7-star tourist lodges in the area: the Sasakwa Hilltop, Sabora plains, and Faru Faru River lodges with a total of 72 beds. The vision of SGRL is rehabilitation and maintenance of the

DOI: http://dx.doi.org/10.5772/intechopen.83827

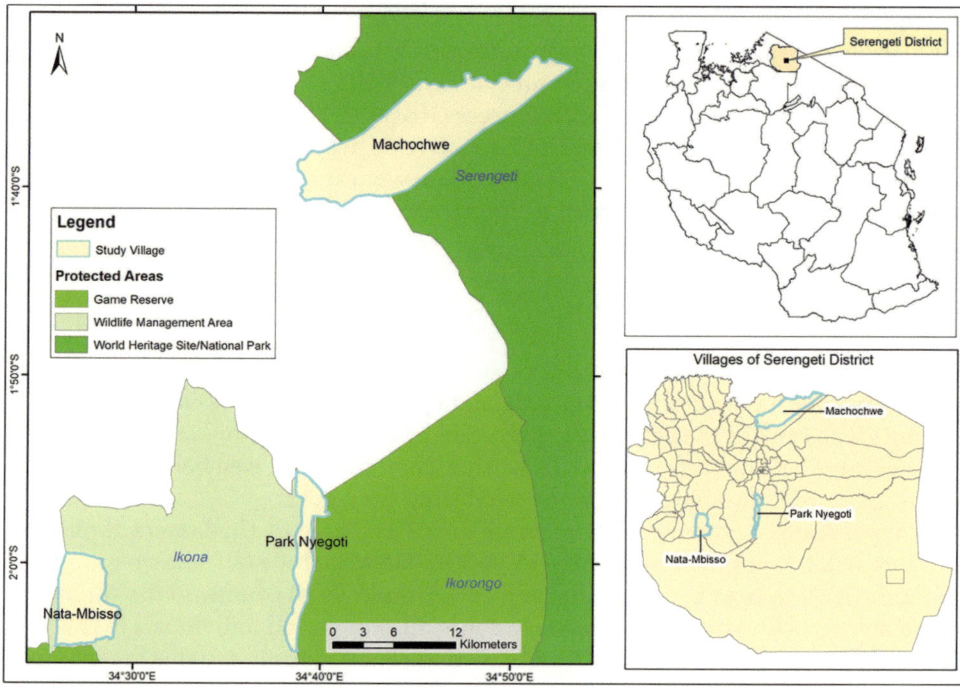

Figure 1.
Location of the study area.

indigenous biodiversity of the western Serengeti-Mara ecosystem for the benefit of the local communities within the Serengeti District and the nation as a whole through structures that are financially and economically sustainable, ecologically and environmentally responsible, and politically acceptable.

From 2003 to 2011 the company had invested a huge sum of money into lodges, community development, and environmental projects in Serengeti and Bunda Districts. Out of this funding, 2.8 and 2.3% have been used for community development projects and wildlife management activities, respectively. Among the community development projects include President Kikwete scholarships, provision of clean and safe water, education, and beekeeping projects.

In 2002 SGRL project entered a joint venture agreement with Serengeti District Council (SDC) on resident hunting ban, and Clause 4.1.6 of the contract states that: "SGRL shall compensate the District Council a sum of TZS 200 Million only for the community development quota allocated to the District Council. The amount will be allocated in a transparent and accountable manner involving all stakeholders in the Serengeti District" (JVA, Pp. 12).

However, SGRL continues with tourist hunting in the former hunting areas incongruent to the joint venture agreement.

3. Methodology

3.1 Socioeconomic data collection

Data involving humans were collected using household questionnaire and structured interviews. The main subjects were households, district officials, and village leaders. Using the village register book, a sample of 30–35 households was picked using simple random sampling technique. Gender balance was considered in picking the household sample. The sample was approximately 7–9% of the village population and can be argued to be an appropriate sample for the study. A total of 100 household respondents were interviewed. As for District officials, officials involved were the District Natural Resources Officer (DNRO) and District Game Officer (DGO). At village level, Village Executive Officers from the three villages were the main respondents. Socioeconomic data were analyzed using SPSS version 16, and this was supplemented with qualitative data analysis techniques such as content analysis and memoing.

3.2 Wildlife population estimates

Between August 16 and August 21 of 2010, the survey of wildlife (large mammals) was undertaken in the census area (data presented are for IKONA WMA). During game census, the census was preceded by a generally above-average wet season rain. The 2 months prior to the census were, however, well below average. No rain was recorded in August and none fell during the census.

Temperatures were generally moderate to warm. Three methods were applied to estimate wildlife populations in IKONA WMA. These include (i) "known groups" method, (ii) total area aerial counts, and (iii) sample aerial counts. In the "known group method," an attempt was made to count all groups and individuals in the population by individual and group recognition. This was only suitable for the rarer species, and so far, it was only attempted for the roan antelope.

For aerial counts using a helicopter (90 m above ground), two forms of aerial counts were implemented simultaneously, namely, total area count and sample estimate based on distance sampling principles [5]. The techniques were implemented following the standard procedures as described by [6]. The air speed was between 40 and 60 knots. Transects were flown morning and afternoon, up to maximum of 3 hours. The hottest part of the day was avoided, as animals tended to rest under the shade at this time and as a consequence are more difficult to spot. Data analysis also followed standard procedures [6].

4. Results and discussion

4.1 Demographic characteristics

On average the majority of the respondents were female which formed 46% (N = 100) except in Natta Mbiso where the majority were males (**Table 1**). As for age about 74% were aged between 18 and 45 years implying that the majority of the sample populations were youths and therefore an economically active workforce.

Village	N	Sex (%)		Age (%)				Education (%)		
		M	F	18–25	26–35	36–45	>45	Inf.	Pry	Sec
Park Nyigoti	35	47.4	52.6	31.6	10.5	21.1	36.9	15.8	68.4	15.8
Natta Mbiso	30	54.5	45.5	27.3	13.6	36.4	22.7	0	63.6	36.4
Machochwe	35	35.5	64.5	17.6	23.5	41.2	17.6	11.8	82.4	5.8
Average	33.3	45.8	54.2	25.5	15.9	32.9	25.7	9.2	71.5	19.3

Table 1.
Demographic data of the sample population.

At specific village level, Machochwe and Natta Mbiso had the highest proportion of youths with 82 and 77%, respectively. In terms of education, the majority (72%) had attained primary school education, 9% have not been to school, and 19% had secondary school education and above. Of the three villages, Natta Mbiso had the highest proportion of people with secondary education. This could be attributed to its strategic location as it is along the Musoma-Arusha main road and the village's sub-township nature compared to the remaining two villages.

4.2 Awareness and people involvement in the hunting ban

On average 95% of household respondents in the three villages were aware of the ban (**Figure 2**). The high awareness could probably be attributed to the fact that these communities have for a long time been linked to bush meat hunting and/or consuming. The fact that they are no longer enjoying access to bush meat as it used to be has probably made them to be aware of the ban. On whether they were involved in meetings and/or discussions before implementing the ban, 95% argued that they were not involved at all, and the remaining proportion were not sure.

However, the District Natural Resources Officer (DNRO) and District Game Officer (DGO), when asked about community involvement, both argued that the decision was reached through the Serengeti District Full Council, and therefore councilors being community representatives had the duty to give feedbacks to them. According to Tanzania administrative setup, each ward should have a councilor who is normally elected by village members (aged >18 years). A Ward may consist of more than three villages.

Local communities were further probed to comment on whether the ban has led to conflicts or not. Results indicate that in Natta Mbiso and Machochwe, 92% had the view that the decision has not led to conflicts (**Figure 3**). However, in Park Nyigoti, 62% argued that the ban has led to increased conflicts. Human-wildlife conflicts in particular between elephants and local communities are widespread in Africa and are a major concern for both elephant conservation and rural development [7]. The increase in conflict in this village could be due to its location. The village is almost an island as it is surrounded by several protected areas (Serengeti National Park, IKONA WMA, and Ikorongo Game Reserve). This has made the village to have very little livelihood options mainly because of very limited land for crop production, increased crop destruction, and livestock predation by wild animals.

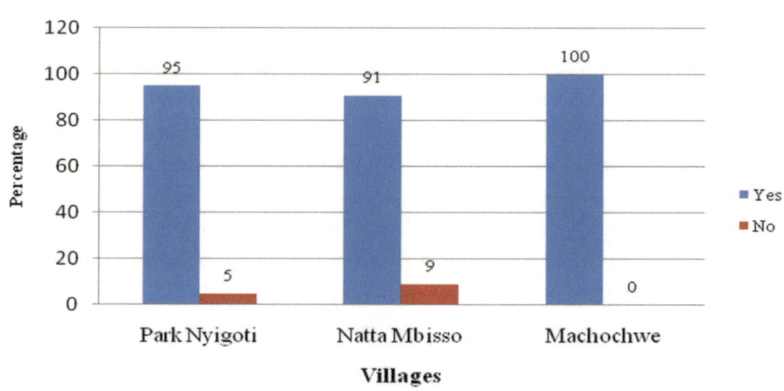

Figure 2.
Awareness on hunting ban.

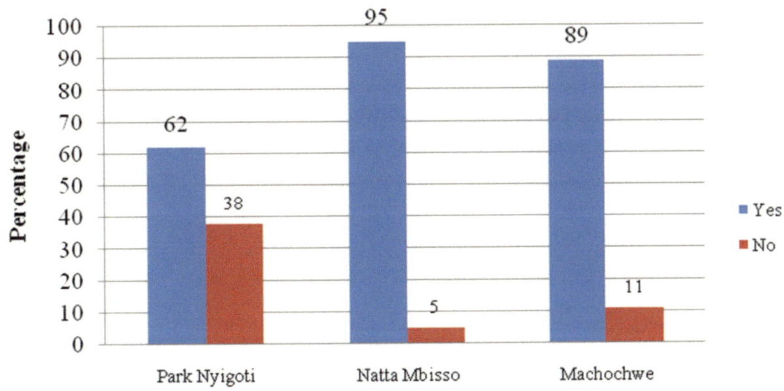

Figure 3.
Ban and increased conflicts.

Park Nyigoti	Natta Mbiso	Machochwe
Construction of dispensary	Employment	Construction of dispensary
Student scholarships	Sell of farm products	
Increased tourists		
Reduced burden on levy		
Construction of deep wells		

Table 2.
Perceived benefits of hunting ban.

Failure to attach conflicts to the ban in Natta Mbiso could be attributed by the socioeconomic benefits local communities are currently enjoying from SGRL. Natta Mbiso currently has a suburban environment as a result of several business enterprises (lodges, hotels, shops, tourist souvenirs, etc.) and the fact that most SGRL employees are settled here. Furthermore, SGRL is currently supporting income-generating enterprises to adjacent communities. As for Machochwe, the possible explanation could be continued poaching as the village boundary is very close to SNP and therefore easy to poach without being easily noticed. In some parts park boundaries are less than a kilometer from the village boundary (Senior Author, pers. Obs.).

4.3 Perceived benefits of hunting ban

Communities had the view that probably the hunting ban has led to increased social services infrastructures (e.g., construction of water wells and dispensary) particularly in Park Nyigoti and Machochwe villages (**Table 2**). The question of what happens to local people is one that remains poorly addressed in conservation literature. This is because it is being produced in an institutional and ideological climate in which there is a widening gap between rhetoric and reality [8]. In such a context [9], it becomes easy to present facile paradigms of how local people will participate in and benefit from conservation interventions.

According to the contract between Serengeti District Council and SGRL, the former is obliged to support social infrastructure developments in the district. SGRL apart from the contribution as per contract also have corporate social responsibility of supporting local communities. In Park Nyigoti other perceived benefits include student bursary, increased tourists, and reduced levy contributions. SGRL

DOI: http://dx.doi.org/10.5772/intechopen.83827

continues to sponsor District Council staff and local communities to pursue different short and long courses within and outside the country. Employment and increased number of tourists were also mentioned as benefits. Selling of farm products such as horticultural crops, meat, eggs, milk, etc. to SGRL staff was also seen as benefit. As for levies, the contribution of SGRL in supporting socioeconomic projects has to a greater extent reduced local community contributions toward development projects. On the other hand, wildlife experts viewed increased revenue to Serengeti District Council (SDC) as a benefit since before the initiative the Council used to receive an average of US $7000/annum (District Game Officer, Pers. Comm).

4.4 Perceived costs of hunting ban

Local community's perceived costs after hunting ban include decreased cash income, food insecurity, malnutrition, increased human-wildlife conflicts, disease transmission, funds' failure to trickle down to primary beneficiaries, and denial of human rights (**Table 3**). Resident hunting has denied village governments and/or local communities to sell game meat, and this has directly denied them from realizing cash income. Local communities used to sell excess bush meat and the former Serengeti Regional Conservation Project (SRCP) had community cropping scheme which made communities to access bush meat at a relatively cheaper price than cattle meat. [10] argue that illegal hunting in the Serengeti has been flourishing, despite stringent law enforcement, because its returns were 45 times greater than those provided legally through the Serengeti Regional Conservation Project community cropping scheme.

Failure to access bush meat has also led to increased protein deficiency in the area. Bush meat was a relatively cheap source of protein, and now most local communities are unable to buy cattle meat due to high price attached to it. During the time of survey, the price of a kg of meat stood at TZS 4000 (US $ 3). This is a relatively high price taking into account the fact that the majority live below a dollar per day. According to [11], per capita income of communities adjacent to Western Serengeti National Park was about US $ 280/annum, an equivalent of US $ 0.77/day.

Increased human-wildlife conflicts were also seen as costs inflicted to communities. The conflicts are of different forms ranging from loss of life due to increased wildlife particularly elephants, transmission of zoonotic diseases to domesticated cattle, and crop destruction by wildlife. Other costs were denial of human rights to access wildlife use contrary to the country's wildlife policy and failure of funds to reach the wider community. The Tanzania Wildlife Policy has the following statement related to wildlife use: "Resident hunting is the right of indigenous Tanzanians

Park Nyigoti	Natta Mbiso	Machochwe
Decreased income	Decreased income	Decrease in income
Increased destructive animals	Increased destructive animals	Increased destructive animals
Protein deficiency/malnutrition	Lack of game meat	Funds do not reach communities
Increased zoonotic diseases	Increased loss of life	Increased loss of life
	Increased poverty	Food insecurity
	Food insecurity	Denial of human rights

Table 3.
Perceived costs of hunting ban.

Year	Amount in USD (US $*)
2004	18,000
2005	75,000
2006	85,000
2007	85,000
2008	71,430
2009	142,850
2010	137,900
2011	133,300
Total	727,000

Source: DGO Office—Serengeti District Council.
**1 USD = TZS 2300; NA = not applicable.*

Table 4.
SGRL revenue contributions 2004–2011.

to have legal access to wildlife use" [4]. During informal discussion with one resident in Natta Mbiso (located close to SGRL) who preferred anonymity had this to say: "SGRL objective is not for conservation and development but to protect her lodges from disturbances (increase visitor attraction) as they own several lodges in the area including her luxurious lodges such as Sasakwa lodge, Sibora-Grumeti lodge, Zebra-Grumeti, Serengeti, Banguesi, Ramahi and Farufaru lodges."

4.5 Compensation realized by the district council

Between 2004 and 2011, the Serengeti District Council received compensation from SGRL amounting to TZS 863 million (US $ 375,217) which in essence was supposed to be used for initiating or improving socioeconomic projects (**Table 4**). Taking into account the fact that the district has more than 50 villages, one can argue that the amount disbursed per village is too little to make an impact on people's livelihood. For example, if the compensation is equally distributed among 50 villages forming the district, each village will get approximately TZS 4 million/annum (US $ 1739) or TZS 11,000/day (US $ 4.8). With the current average household size of six people in the district, each household is likely to receive about TZS 1833/day (US $ 0.8). By all standards this is very little money for any meaningful poverty reduction contribution at community level. The game species commonly hunted by residents are mainly impala, Thomson's gazelle, wildebeest, topi, eland, buffalo, bohor reedbuck, and Grant's gazelle. Other wildlife species counted during the survey were the elephant, giraffe, hartebeest, and zebra.

5. Population trends for selected wildlife species

The population trends for selected wildlife species in IKONA WMA, that is, impala, Thomson's gazelle, wildebeest, buffalo, and eland, are presented in **Figures 4–8**. These species are preferred for bush meat.

Results indicate that for the first 2–5 years after the ban, the population showed an increasing trend before declining for impala, Thomson's gazelle, wildebeest, buffalo, and eland (**Figures 4–7**). The increase for the first 5 years before recording a decline could be attributed to fewer disturbances and/or the presence of conducive environment to wildlife. Hunting activities normally disrupt wildlife

DOI: http://dx.doi.org/10.5772/intechopen.83827

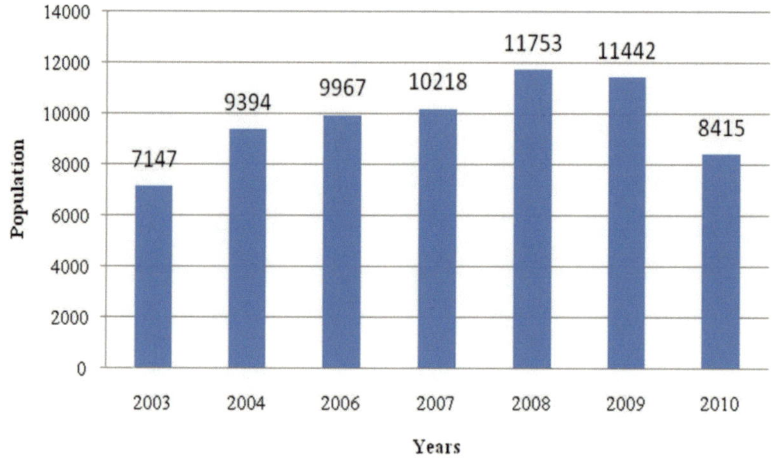

Figure 4.
Population trend of impala.

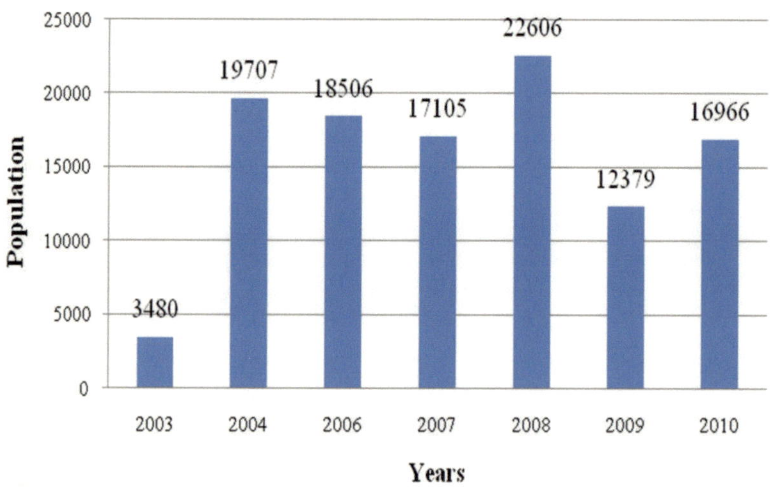

Figure 5.
Population trend of Thomson's gazelle. Source: Goodman [6].

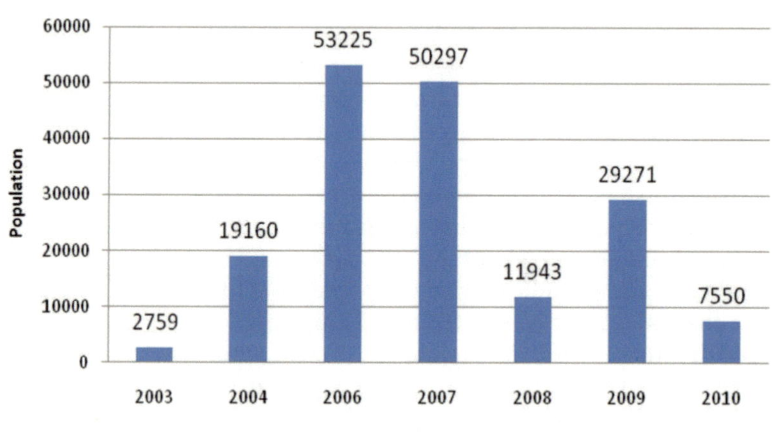

Figure 6.
Population trend of wildebeest.

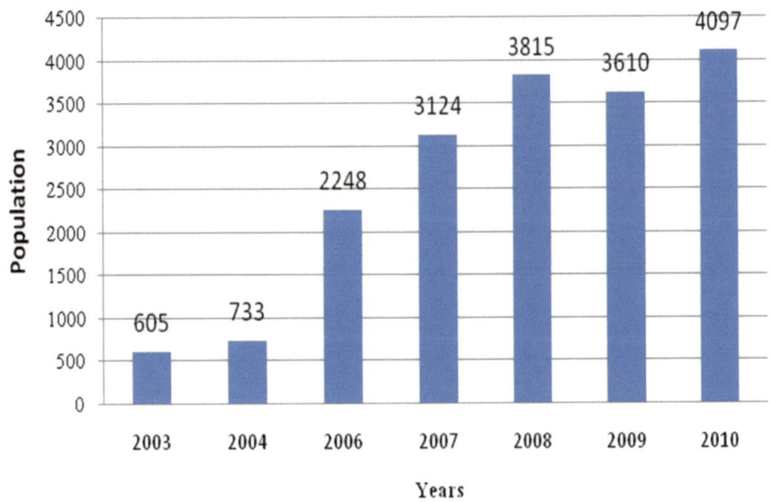

Figure 7.
Population trend of buffalo.

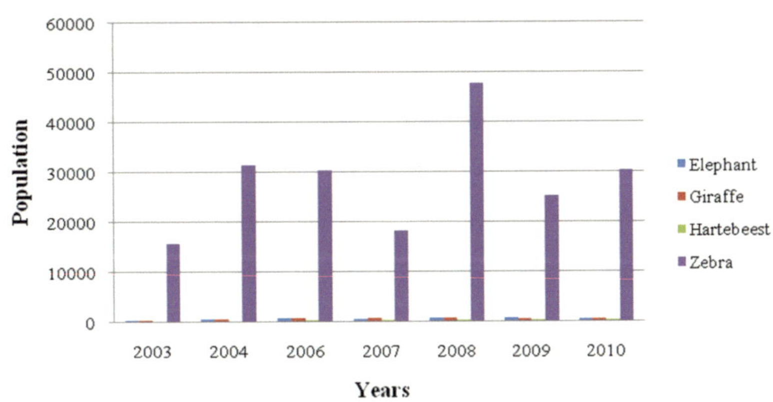

Figure 8.
Population trends for species not in hunting quota. Source: Goodman [6].

ecological systems and make wildlife timid and less productive. The reasons for the decline after the fifth year could not be established. However, this can be attributed to poaching as illustrated in **Table 5**. According to [12, 13], incidences of wildlife crimes and/or animal killed led the wildlife to flee to nearby protected areas and/or distant protected areas particularly for wide-ranging animals such as elephants. The discussion with one elder who preferred anonymity had this to say: "after the ban communities have now intensified poaching to compensate for the previous opportunity of getting bush meat at a cheaper price."

Population trends for other wildlife species (normally not included in hunting quota in the area), namely, elephant, giraffe, zebra, and hartebeest, are presented in **Figure 8**. Field findings indicate that the population for zebra fluctuated over time, but for the rest (elephant, giraffe, and hartebeest), the population remained constant. For the first 2 years, the population of zebra increased before dropping for 2 years and then rose to the maximum (about 5000) in 2008. It then dropped and rises again. The rise in the first 2 years can be attributed to the safe environment after the ban and availability of habitat. The decline could be associated with hunting for commercial and subsistence. In the study area, zebra meat is highly preferred bush meat by the community due to its deliciousness (Magoiga, pers. comm.).

Year	Wildlife crime incidents	Animals killed	Traditional weapon
2003	147	238	1573
2004	330	824	4826
2005	342	325	3295
2006	481	1387	6160
2007	353	524	371
2008	300	395	2019
2009	315	278	1571
2010	271	406	2792
2011	361	216	2346
2012	238	213	848
2013	266	152	1884

Source: Mwakalobo et al. [12].

Table 5.
Trend of wildlife crime incidents, animals killed, and traditional weapons caught.

6. Conclusion and recommendations

The study concludes that the ban has significantly boosted the District Council revenues. Despite this positive development, the livelihood of the communities has been significantly affected in many ways including increased human-wildlife conflicts, food insecurity, malnutrition, and lack of bush meat which to many is regarded as cheap source of protein. In addition, the ban led to the increase of wildlife populations particularly in the first 2–5 years. This increase corresponded with increased human-wildlife conflicts as a result of crop destruction and livestock predation. The study recommends the need for involvement of communities in major decisions affecting people's livelihood such as hunting ban. We also recommend that before instituting hunting ban(s), this has to be informed by research.

Acknowledgements

We are thankful to the Department of Geography and Environmental Studies, University of Dodoma, for the permission to collect data for this study.

Conflict of interest

Authors declare no conflict of interest.

Author details

Abiud L. Kaswamila* and Augustino E. Mwakipesile
Department of Geography and Environmental Studies, The University of Dodoma, Dodoma, Tanzania

*Address all correspondence to: abagore.kaswamila6@gmail.com

DOI: http://dx.doi.org/10.5772/intechopen.83827

References

[1] Wall B, Child B. When Does Hunting Contribute to Conservation and Rural Development? In: Barney D, Hutton J, William MA, editors. 2009. pp. 253-265

[2] Mahoney SP. Recreational Hunting and Sustainable Use in North America; 2009. pp. 266-281. DOI: 10.1002/9781444303179. Ch. 16

[3] Wall WA, Kernohan B. Key components of conservation-hunting programmes and their relationships to populations, ecosystems, and people. In: Paper presented at the 3rd International Wildlife Management Congress; 1-5 December 2003; Christchurch, New Zealand

[4] URT. The Wildlife Policy of Tanzania. Dar es Salaam: Government Printer; 2007

[5] Buckland ST, Anderson DD, Burnham KP, Laake JL. Distance Sampling. Estimating Abundance of Biological Populations. 1st ed. London: Chapman and Hall; 1993

[6] Goodman PS. Large Herbivore Population Estimates for Grumeti Reserves, Unpublished Report. Sasakwa, Tanzania: Grumeti Fund; 2011

[7] Woodroffe R, Thirgood S, Rabinowitz A. The impact of human-wildlife conflict on human lives and livelihoods. In: Woodroffe R, Thirgood S, Rabinowitz A, editors. People and Wildlife; Conflict or Coexistence? Conservation Biology 9. Cambridge; 2005. pp. 13-26

[8] Leader-Williams N, Hutton MJ. Does extractive use provide opportunities to offset conflicts between people and wildlife? The impact of human-wildlife conflict on human lives and livelihoods. In: Woodroffe R, Thirgood S, Rabinowitz A, editors. People and Wildlife; Conflict or Coexistence? Conservation Biology 9. Cambridge; 2005. pp. 140-161

[9] Brockington D, Igoe J. Eviction for conservation: A global overview. Conservation and Society. 2007;**4**(3):424-471

[10] Holmern T, Roskaft E, Mbaruku J, Mkama S, Muya J. Uneconomical game cropping in a community-based conservation project outside Serengeti national, Tanzania. Oryx. 2002;**36**:364-372

[11] Kaswamila AL. Impact of wildlife on household food security and income in northern Tanzania, Journal of Human Dimensions of Wildlife. 2007;**12**:331-352

[12] Mwakalobo A, Kaswamila AL, Kira A, Chawala O, Tear T. Tourism regional multiplier effects in Tanzania: Analysis of Singita Grumeti reserves tourism in Mara region. Journal of Sustainable Development. 2016;**9**(4):40-60

[13] United Republic of Tanzania. Mara Region Basic Demographic and Socio Economic Profile. National Bureau of Statistics; 2016